the Theory and
Technique of
Electronic
Music

Miller Puckette

University of California, San Diego, USA

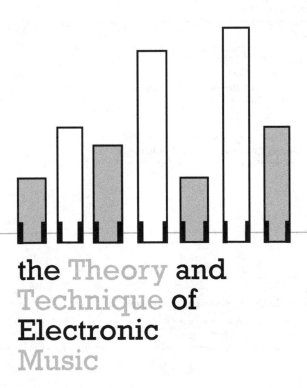

the Theory and
Technique of
Electronic
Music

 **World Scientific**

NEW JERSEY • LONDON • SINGAPORE • BEIJING • SHANGHAI • HONG KONG • TAIPEI • CHENNAI

Published by

World Scientific Publishing Co. Pte. Ltd.

5 Toh Tuck Link, Singapore 596224

USA office: 27 Warren Street, Suite 401-402, Hackensack, NJ 07601

UK office: 57 Shelton Street, Covent Garden, London WC2H 9HE

Library of Congress Cataloging-in-Publication Data
Puckette, Miller, 1959–
    The theory and technique of electronic music / Miller Puckette.
      p. cm.
    Includes bibliographical references and index.
    ISBN-13 978-981-270-077-3
    ISBN-10 981-270-077-3
    1. Electronic music--Instruction and study.
    ML1380 .P83 2007

                2008295011

British Library Cataloguing-in-Publication Data
A catalogue record for this book is available from the British Library.

First published 2007
Reprinted 2010, 2011

Printed in Singapore.

Contents

Foreword

The Theory and Technique of Electronic Music is a uniquely complete source of information for the computer synthesis of rich and interesting musical timbres. The theory is clearly presented in a completely general form. But in addition, examples of how to synthesize each theoretical aspect are presented in the Pd language so the reader of the book can immediately use the theory for his musical purposes. I know of no other book which combines theory and technique so usefully.

By far the most popular music and sound synthesis programs in use today are block diagram compilers with graphical interfaces. These allow the composer to design instruments by displaying the "objects" of his instrument on a computer screen and drawing the connecting paths between the objects. The resulting graphical display is very congenial to musicians. A naive user can design a simple instrument instantly. He can rapidly learn to design complex instruments. He can understand how complex instruments work by looking at their graphical images.

The first graphical compiler program, Max, was written by Miller Puckette in 1988. Max dealt only with control signals for music synthesis because the computers available at the time were not fast enough to deal with sound. As soon as faster computers which could compute soundwave samples in real-time were available, Puckette and David Zicarelli appended MSP to Max (Max/MSP) thus making the computer, usually a laptop computer, into a complete musical instrument capable of live performance.

Development of Max/MSP was done by Puckette and Zicarelli at IR-CAM in the period 1993 to 1994 . Both have now moved to California. Zicarelli commercialized and sells Max, MSP, and JITTER (an extension to video synthesis) as products. Puckette, now a professor at UCSD, wrote Pd (Pure Data). It is an open source program which is a close equivalent to Max/MSP.

Max and Pd allow almost anyone to synthesize uninteresting timbres almost instantly. Making interesting timbres is much more difficult and requires much additional knowledge. *The Theory and Technique of Elec-*

tronic Music is that body of knowledge. The theory is important for any synthesis program. *The Theory and Technique of Electronic Music* gives copious examples of how to apply the theory using Pd. The combination of theory plus Pd examples makes this book uniquely useful. It also contains problem sets for each chapter so it is a fine textbook.

I expect Puckette's book to become THE essential book in any electronic musician's library.

Max Mathews

Preface

This is a book about using electronic techniques to record, synthesize, process, and analyze musical sounds, a practice which came into its modern form in the years 1948-1952, but whose technological means and artistic uses have undergone several revolutions since then. Nowadays most electronic music is made using computers, and this book will focus exclusively on what used to be called "computer music", but which should really now be called "electronic music using a computer".

Most of the computer music tools available today have antecedents in earlier generations of equipment. The computer, however, is relatively cheap and the results of using one are easy to document and re-create. In these respects at least, the computer makes the ideal electronic music instrument—it is hard to see what future technology could displace it.

The techniques and practices of electronic music can be studied (at least in theory) without making explicit reference to the current state of technology. Still, it's important to provide working examples. So each chapter starts with theory (avoiding any reference to implementation) and ends with a series of examples realized in a currently available software package.

The ideal reader of this book is anyone who knows and likes electronic music of any genre, has plenty of facility with computers in general, and who wants to learn how to make electronic music from the ground up, starting with the humble oscillator and continuing through sampling, FM, filtering, waveshaping, delays, and so on. This will take plenty of time.

This book doesn't take the easy route of recommending pre-cooked software to try out these techniques; instead, the emphasis is on learning how to use a general-purpose computer music environment to realize them yourself. Of the several such packages available, we'll use Pd, but that shouldn't stop you from using these same techniques in other environments such as Csound or Max/MSP.

To read this book you must understand mathematics through intermediate algebra and trigonometry; starting in Chapter 7, complex numbers

xiii

also make an appearance, although not complex analyis. (For instance, complex numbers are added, multiplied, and conjugated, but there are no complex exponentials.) A review of mathematics for computer music by F. Richard Moore appears in [Str85, pp. 1-68].

Although the "level" of mathematics is not high, the mathematics itself is sometimes quite challenging. All sorts of cool mathematics is in the reach of any student of algebra or geometry. In the service of computer music, for instance, we'll run into Bessel functions, Chebychev polynomials, the Central Limit Theorem, and, of course, Fourier analysis.

You don't need much background in music as it is taught in the West; in particular, Western written music notation is not needed. Some elementary bits of Western music theory are used, such as the tempered scale, the A-B-C system of naming pitches, and terms like "note" and "chord". Also you should be familiar with terms of musical acoustics such as sinusoids, amplitude, frequency, and the overtone series.

Each chapter starts with a theoretical discussion of some family of techniques or theoretical issues, followed by a series of examples realized in Pd to illustrate them. The examples are included in the Pd distribution, so you can run them and/or edit them into your own spinoffs. In addition, all the figures were created using Pd patches, which appear in an electronic supplement. These aren't carefully documented but in principle could be used as an example of Pd's drawing capabilities for anyone interested in that.

I would like to thank some people who have made it possible for me to write this. Barry Vercoe is almost entirely responsible for my music education. Meanwhile I was taught mathematics by Wayne Holman, Samuel Greitzer, Murray Klamkin, Gian-Carlo Rota, Frank Morgan, Michael Artin, Andrew Gleason, and many others. Phil White taught me English and Rosie Paschall visual composition. Finally, my parents (one deceased) are mighty patient; I'm now 47. Thank you.

Chapter 1

Sinusoids, Amplitude and Frequency

Electronic music is usually made using a computer, by synthesizing or processing *digital audio signals*. These are sequences of numbers,

$$\dots, x[n-1], x[n], x[n+1], \dots$$

where the index n, called the *sample number*, may range over some or all the integers. A single number in the sequence is called a *sample*. An example of a digital audio signal is the *Sinusoid*:

$$x[n] = a\cos(\omega n + \phi)$$

where a is the *amplitude,* ω is the *angular frequency,* and ϕ is the initial *phase.* The phase is a function of the sample number n, equal to $\omega n + \phi$. The initial phase is the phase at the zeroth sample ($n = 0$).

Figure 1.1 (part a) shows a sinusoid graphically. The horizontal axis shows successive values of n and the vertical axis shows the corresponding values of $x[n]$. The graph is drawn in such a way as to emphasize the sampled nature of the signal. Alternatively, we could draw it more simply as a continuous curve (part b). The upper drawing is the most faithful representation of the (digital audio) sinusoid, whereas the lower one can be considered an idealization of it.

Sinusoids play a key role in audio processing because, if you shift one of them left or right by any number of samples, you get another one. This makes it easy to calculate the effect of all sorts of operations on sinusoids. Our ears use this same special property to help us parse incoming

1

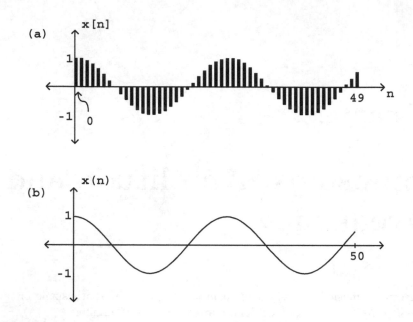

Figure 1.1: A digital audio signal, showing its discrete-time nature (part a), and idealized as a continuous function (part b). This signal is a (real-valued) sinusoid, fifty points long, with amplitude 1, angular frequency 0.24, and initial phase zero.

sounds, which is why sinusoids, and combinations of sinusoids, can be used to achieve many musical effects.

Digital audio signals do not have any intrinsic relationship with time, but to listen to them we must choose a *sample rate*, usually given the variable name R, which is the number of samples that fit into a second. The time t is related to the sample number n by $Rt = n$, or $t = n/R$. A sinusoidal signal with angular frequency ω has a real-time frequency equal to

$$f = \frac{\omega R}{2\pi}$$

in Hertz (i.e., cycles per second), because a cycle is 2π radians and a second is R samples.

A real-world audio signal's amplitude might be expressed as a time-varying voltage or air pressure, but the samples of a digital audio signal are unitless numbers. We'll casually assume here that there is ample numerical accuracy so that we can ignore round-off errors, and that the numerical format is unlimited in range, so that samples may take any value we wish.

However, most digital audio hardware works only over a fixed range of input and output values, most often between -1 and 1. Modern digital audio processing software usually uses a floating-point representation for signals. This allows us to use whatever units are most convenient for any given task, as long as the final audio output is within the hardware's range [Mat69, pp. 4-10].

1.1 Measures of Amplitude

The most fundamental property of a digital audio signal is its amplitude. Unfortunately, a signal's amplitude has no one canonical definition. Strictly speaking, all the samples in a digital audio signal are themselves amplitudes, and we also spoke of the amplitude a of the sinusoid as a whole. It is useful to have measures of amplitude for digital audio signals in general. Amplitude is best thought of as applying to a *window*, a fixed range of samples of the signal. For instance, the window starting at sample M of length N of an audio signal $x[n]$ consists of the samples,

$$x[M], x[M+1], \ldots, x[M+N-1]$$

The two most frequently used measures of amplitude are the *peak amplitude*, which is simply the greatest sample (in absolute value) over the window:

$$A_{\text{peak}}\{x[n]\} = \max |x[n]|, \qquad n = M, \ldots, M+N-1$$

and the *root mean square* (RMS) amplitude:

$$A_{\text{RMS}}\{x[n]\} = \sqrt{P\{x[n]\}}$$

where $P\{x[n]\}$ is the mean *power*, defined as:

$$P\{x[n]\} = \frac{1}{N}\left(|x[M]|^2 + \cdots + |x[M+N-1]|^2\right)$$

(In this last formula, the absolute value signs aren't necessary at the moment since we're working on real-valued signals, but they will become important later when we consider complex-valued signals.) Neither the peak nor the RMS amplitude of any signal can be negative, and either one can be exactly zero only if the signal itself is zero for all n in the window.

The RMS amplitude of a signal may equal the peak amplitude but never exceeds it; and it may be as little as $1/\sqrt{N}$ times the peak amplitude, but never less than that.

Under reasonable conditions—if the window contains at least several periods and if the angular frequency is well under one radian per sample—the peak amplitude of the sinusoid of Page 1 is approximately a and its RMS

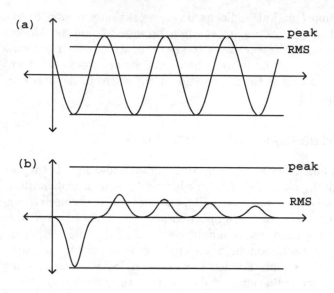

Figure 1.2: Root mean square (RMS) and peak amplitudes of signals compared. For a sinusoid (part a), the peak amplitude is higher than RMS by a factor of $\sqrt{2}$.

amplitude about $a/\sqrt{2}$. Figure 1.2 shows the peak and RMS amplitudes of two digital audio signals.

1.2   Units of Amplitude

Two amplitudes are often better compared using their ratio than their difference. Saying that one signal's amplitude is greater than another's by a factor of two might be more informative than saying it is greater by 30 millivolts. This is true for any measure of amplitude (RMS or peak, for instance). To facilitate comparisons, we often express amplitudes in logarithmic units called *decibels*. If a is the amplitude of a signal (either peak or RMS), then we can define the decibel (dB) level d as:

$$d = 20 \cdot \log_{10}(a/a_0)$$

where a_0 is a reference amplitude. This definition is set up so that, if we increase the signal power by a factor of ten (so that the amplitude increases by a factor of $\sqrt{10}$), the logarithm will increase by $1/2$, and so the value in

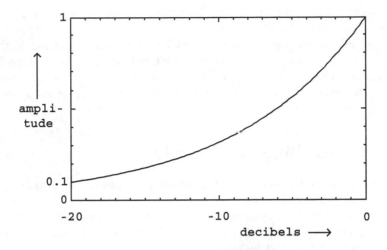

Figure 1.3: The relationship between decibel and linear scales of amplitude. The linear amplitude 1 is assigned to 0 dB.

decibels goes up (additively) by ten. An increase in amplitude by a factor of two corresponds to an increase of about 6.02 decibels; doubling power is an increase of 3.01 dB. The relationship between linear amplitude and amplitude in decibels is graphed in Figure 1.3.

Still using a_0 to denote the reference amplitude, a signal with linear amplitude smaller than a_0 will have a negative amplitude in decibels: $a_0/10$ gives -20 dB, $a_0/100$ gives -40, and so on. A linear amplitude of zero is smaller than that of any value in dB, so we give it a dB level of $-\infty$.

In digital audio a convenient choice of reference, assuming the hardware has a maximum amplitude of one, is

$$a_0 = 10^{-5} = 0.00001$$

so that the maximum amplitude possible is 100 dB, and 0 dB is likely to be inaudibly quiet at any reasonable listening level. Conveniently enough, the dynamic range of human hearing—the ratio between a damagingly loud sound and an inaudibly quiet one—is about 100 dB.

Amplitude is related in an inexact way to the perceived loudness of a sound. In general, two signals with the same peak or RMS amplitude won't necessarily have the same loudness at all. But amplifying a signal by 3 dB, say, will fairly reliably make it sound about one "step" louder. Much has been made of the supposedly logarithmic nature of human hearing (and other senses), which may partially explain why decibels are such a useful scale of amplitude [RMW02, p. 99].

Amplitude is also related in an inexact way to musical *dynamic*. Dynamic is better thought of as a measure of effort than of loudness or power. It ranges over nine values: rest, ppp, pp, p, mp, mf, f, ff, fff. These correlate in an even looser way with the amplitude of a signal than does loudness [RMW02, pp. 110-111].

1.3   Controlling Amplitude

Perhaps the most frequently used operation on electronic sounds is to change their amplitudes. For example, a simple strategy for synthesizing sounds is by combining sinusoids, which can be generated by evaluating the formula on Page 1, sample by sample. But the sinusoid has a constant nominal amplitude a, and we would like to be able to vary that in time.

In general, to multiply the amplitude of a signal $x[n]$ by a factor $y \geq 0$, you can just multiply each sample by y, giving a new signal $y \cdot x[n]$. Any measurement of the RMS or peak amplitude of $x[n]$ will be greater or less by the factor y. More generally, you can change the amplitude by an amount $y[n]$ which varies sample by sample. If $y[n]$ is nonnegative and if it varies slowly enough, the amplitude of the product $y[n] \cdot x[n]$ (in a fixed window from M to $M + N - 1$) will be that of $x[n]$, multiplied by the value of $y[n]$ in the window (which we assume doesn't change much over the N samples in the window).

In the more general case where both $x[n]$ and $y[n]$ are allowed to take negative and positive values and/or to change quickly, the effect of multiplying them can't be described as simply changing the amplitude of one of them; this is considered later in Chapter 5.

1.4   Frequency

Frequencies, like amplitudes, are often measured on a logarithmic scale, in order to emphasize proportions between them, which usually provide a better description of the relationship between frequencies than do differences between them. The frequency ratio between two musical tones determines the musical interval between them.

The Western musical scale divides the *octave* (the musical interval associated with a ratio of 2:1) into twelve equal sub-intervals, each of which therefore corresponds to a ratio of $2^{1/12}$. For historical reasons this sub-interval is called a *half-step*. A convenient logarithmic scale for pitch is simply to count the number of half-steps from a reference pitch—allowing fractions to permit us to specify pitches which don't fall on a note of the Western scale. The most commonly used logarithmic pitch scale is "MIDI pitch", in which the pitch 69 is assigned to a frequency of 440 cycles per second—the A above middle C. To convert between a MIDI pitch m and a frequency in cycles per second f, apply the Pitch/Frequency Conversion formulas:

$$m = 69 + 12 \cdot \log_2(f/440)$$

$$f = 440 \cdot 2^{(m-69)/12}$$

Middle C, corresponding to MIDI pitch $m = 60$, comes to $f = 261.626$ cycles per second.

MIDI itself is an old hardware protocol which has unfortunately insinuated itself into a great deal of software design. In hardware, MIDI allows only integer pitches between 0 and 127. However, the underlying scale is well defined for any "MIDI" number, even negative ones; for example a "MIDI pitch" of -4 is a decent rate of vibrato. The pitch scale cannot, however, describe frequencies less than or equal to zero cycles per second. (For a clear description of MIDI, its capabilities and limitations, see [Bal03, ch.6-8]).

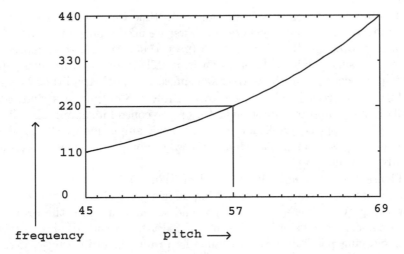

Figure 1.4: The relationship between "MIDI" pitch and frequency in cycles per second (Hertz). The span of 24 MIDI values on the horizontal axis represents two octaves, over which the frequency increases by a factor of four.

A half-step comes to a ratio of about 1.059 to 1, or about a six percent increase in frequency. Half-steps are further divided into *cents*, each cent being one hundredth of a half-step. As a rule of thumb, it might take about three cents to make a discernible change in the pitch of a musical tone. At middle C this comes to a difference of about 1/2 cycle per second. A graph of frequency as a function of MIDI pitch, over a two-octave range, is shown in Figure 1.4.

1.5   Synthesizing a Sinusoid

In most widely used audio synthesis and processing packages (Csound, Max/MSP, and Pd, for instance), the audio operations are specified as networks of *unit generators*[Mat69] which pass audio signals among themselves. The user of the software package specifies the network, sometimes called a *patch*, which essentially corresponds to the synthesis algorithm to be used, and then worries about how to control the various unit generators in time. In this section, we'll use abstract block diagrams to describe patches, but in the "examples" section (Page 18), we'll choose a specific implementation environment and show some of the software-dependent details.

To show how to produce a sinusoid with time-varying amplitude we'll need to introduce two unit generators. First we need a pure sinusoid which is made with an *oscillator*. Figure 1.5 (part a) shows a pictorial representation of a sinusoidal oscillator as an icon. The input is a frequency (in cycles per second), and the output is a sinusoid of peak amplitude one.

Figure 1.5 (part b) shows how to multiply the output of a sinusoidal oscillator by an appropriate scale factor $y[n]$ to control its amplitude. Since the oscillator's peak amplitude is 1, the peak amplitude of the product is about $y[n]$, assuming $y[n]$ changes slowly enough and doesn't become negative in value.

Figure 1.6 shows how the sinusoid of Figure 1.1 is affected by amplitude change by two different controlling signals $y[n]$. The controlling signal shown in part (a) has a discontinuity, and so therefore does the resulting amplitude-controlled sinusoid shown in (b). Parts (c) and (d) show a more gently-varying possibility for $y[n]$ and the result. Intuition suggests that the result shown in (b) won't sound like an amplitude-varying sinusoid, but instead like a sinusoid interrupted by an audible "pop" after which it continues more quietly. In general, for reasons that can't be explained in this chapter, amplitude control signals $y[n]$ which ramp smoothly from one value to another are less likely to give rise to parasitic results (such as that "pop") than are abruptly changing ones.

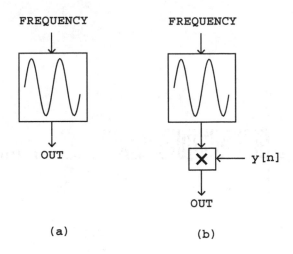

Figure 1.5: Block diagrams for (a) a sinusoidal oscillator; (b) controlling the amplitude using a multiplier and an amplitude signal $y[n]$.

For now we can state two general rules without justifying them. First, pure sinusoids are the signals most sensitive to the parasitic effects of quick amplitude change. So when you want to test an amplitude transition, if it works for sinusoids it will probably work for other signals as well. Second, depending on the signal whose amplitude you are changing, the amplitude control will need between 0 and 30 milliseconds of "ramp" time—zero for the most forgiving signals (such as white noise), and 30 for the least (such as a sinusoid). All this also depends in a complicated way on listening levels and the acoustic context.

Suitable amplitude control functions $y[n]$ may be made using an *envelope generator*. Figure 1.7 shows a network in which an envelope generator is used to control the amplitude of an oscillator. Envelope generators vary widely in design, but we will focus on the simplest kind, which generates line segments as shown in Figure 1.6 (part c). If a line segment is specified to ramp between two output values a and b over N samples starting at sample number M, the output is:

$$y[n] = a + (b - a)\frac{n - M}{N}, \quad M \le n \le M + N - 1$$

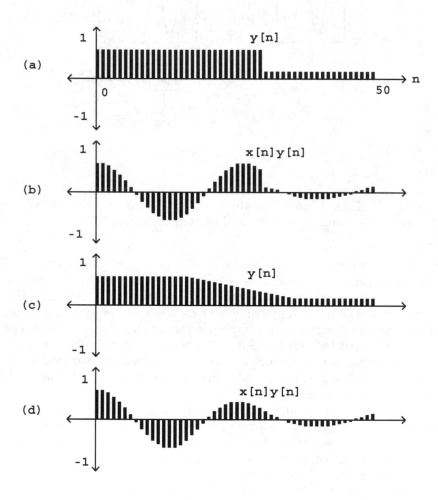

Figure 1.6: Two amplitude functions (parts a, c), and (parts b, d), the result of multiplying them by the pure sinusoid of Figure 1.1.

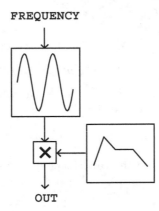

Figure 1.7: Using an envelope generator to control amplitude.

The output may have any number of segments such as this, laid end to end, over the entire range of sample numbers n; flat, horizontal segments can be made by setting $a = b$.

In addition to changing amplitudes of sounds, amplitude control is often used, especially in real-time applications, simply to turn sounds on and off: to turn one off, ramp the amplitude smoothly to zero. Most software synthesis packages also provide ways to actually stop modules from computing samples at all, but here we'll use amplitude control instead.

The envelope generator dates from the analog era [Str95, p.64] [Cha80, p.90], as does the rest of Figure 1.7; oscillators with controllable frequency were called voltage-controlled oscillators or VCOs, and the multiplication step was done using a voltage-controlled amplifier or VCA [Str95, pp.34-35] [Cha80, pp.84-89]. Envelope generators are described in more detail in Section 4.1.

1.6 Superposing Signals

If a signal $x[n]$ has a peak or RMS amplitude A (in some fixed window), then the scaled signal $k \cdot x[n]$ (where $k \geq 0$) has amplitude kA. The mean power of the scaled signal changes by a factor of k^2. The situation gets more complicated when two different signals are added together; just knowing the amplitudes of the two does not suffice to know the amplitude of the sum. The two amplitude measures do at least obey triangle inequalities; for any two signals $x[n]$ and $y[n]$,

$$A_{\text{peak}}\{x[n]\} + A_{\text{peak}}\{y[n]\} \geq A_{\text{peak}}\{x[n] + y[n]\}$$

$$A_{\text{RMS}}\{x[n]\} + A_{\text{RMS}}\{y[n]\} \geq A_{\text{RMS}}\{x[n] + y[n]\}$$

If we fix a window from M to $N + M - 1$ as usual, we can write out the mean power of the sum of two signals:

$$P\{x[n] + y[n]\} = P\{x[n]\} + P\{y[n]\} + 2 \cdot \text{COV}\{x[n], y[n]\}$$

where we have introduced the *covariance* of two signals:

$$\text{COV}\{x[n], y[n]\} = \frac{x[M]y[M] + \cdots + x[M + N - 1]y[M + N - 1]}{N}$$

The covariance may be positive, zero, or negative. Over a sufficiently large window, the covariance of two sinusoids with different frequencies is negligible compared to the mean power. Two signals which have no covariance are called *uncorrelated* (the correlation is the covariance normalized to lie between -1 and 1). In general, for two uncorrelated signals, the power of the sum is the sum of the powers:

$$P\{x[n] + y[n]\} = P\{x[n]\} + P\{y[n]\}, \quad \text{whenever } \text{COV}\{x[n], y[n]\} = 0$$

Put in terms of amplitude, this becomes:

$$\left(A_{\text{RMS}}\{x[n] + y[n]\}\right)^2 = \left(A_{\text{RMS}}\{x[n]\}\right)^2 + \left(A_{\text{RMS}}\{y[n]\}\right)^2$$

This is the familiar Pythagorean relation. So uncorrelated signals can be thought of as vectors at right angles to each other; positively correlated ones as having an acute angle between them, and negatively correlated as having an obtuse angle between them.

For example, if two uncorrelated signals both have RMS amplitude a, the sum will have RMS amplitude $\sqrt{2}a$. On the other hand if the two signals happen to be equal—the most correlated possible—the sum will have amplitude $2a$, which is the maximum allowed by the triangle inequality.

1.7   Periodic Signals

A signal $x[n]$ is said to repeat at a period τ if

$$x[n + \tau] = x[n]$$

for all n. Such a signal would also repeat at periods 2τ and so on; the smallest τ if any at which a signal repeats is called the signal's *period*. In discussing periods of digital audio signals, we quickly run into the difficulty of describing signals whose "period" isn't an integer, so that the equation above doesn't make sense. For now we'll effectively ignore this difficulty by supposing that the signal $x[n]$ may somehow be interpolated between the samples so that it's well defined whether n is an integer or not.

A sinusoid has a period (in samples) of $2\pi/\omega$ where ω is the angular frequency. More generally, any sum of sinusoids with frequencies $2\pi k/\omega$, for integers k, will repeat after $2\pi/\omega$ samples. Such a sum is called a *Fourier Series*:

$$x[n] = a_0 + a_1 \cos(\omega n + \phi_1) + a_2 \cos(2\omega n + \phi_2) + \cdots + a_p \cos(p\omega n + \phi_p)$$

Moreover, if we make certain technical assumptions (in effect that signals only contain frequencies up to a finite bound), we can represent any periodic signal as such a sum. This is the discrete-time variant of Fourier analysis which will reappear in Chapter 9.

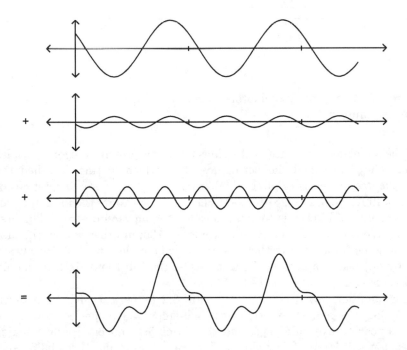

Figure 1.8: A Fourier series, showing three sinusoids and their sum. The three component sinusoids have frequencies in the ratio 1:2:3.

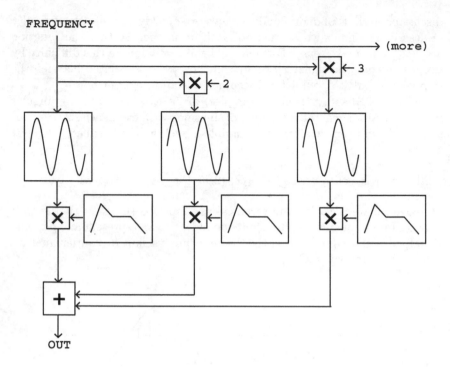

Figure 1.9: Using many oscillators to synthesize a waveform with desired harmonic amplitudes.

The angular frequencies of the sinusoids above are all integer multiples of ω. They are called the *harmonics* of ω, which in turn is called the *fundamental*. In terms of pitch, the harmonics $\omega, 2\omega, \ldots$ are at intervals of 0, 1200, 1902, 2400, 2786, 3102, 3369, 3600, ..., cents above the fundamental; this sequence of pitches is sometimes called the *harmonic series*. The first six of these are all quite close to multiples of 100; in other words, the first six harmonics of a pitch in the Western scale land close to (but not always exactly on) other pitches of the same scale; the third and sixth miss only by 2 cents and the fifth misses by 14.

Put another way, the frequency ratio 3:2 (a perfect fifth in Western terminology) is almost exactly seven half-steps, 4:3 (a perfect fourth) is just as near to five half-steps, and the ratios 5:4 and 6:5 (perfect major and minor thirds) are fairly close to intervals of four and three half-steps, respectively.

A Fourier series (with only three terms) is shown in Figure 1.8. The first three graphs are of sinusoids, whose frequencies are in a 1:2:3 ratio.

The common period is marked on the horizontal axis. Each sinusoid has a different amplitude and initial phase. The sum of the three, at bottom, is not a sinusoid, but it still maintains the periodicity shared by the three component sinusoids.

Leaving questions of phase aside, we can use a bank of sinusoidal oscillators to synthesize periodic tones, or even to morph smoothly through a succession of periodic tones, by specifying the fundamental frequency and the (possibly time-varying) amplitudes of the partials. Figure 1.9 shows a block diagram for doing this.

This is an example of *additive synthesis*; more generally the term can be applied to networks in which the frequencies of the oscillators are independently controllable. The early days of computer music rang with the sound of additive synthesis.

1.8   About the Software Examples

The examples for this book use Pure Data (Pd), and to understand them you will have to learn at least something about Pd itself. Pd is an environment for quickly realizing computer music applications, primarily intended for live music performances. Pd can be used for other media as well, but we won't go into that here.

Several other patchable audio DSP environments exist besides Pd. The most widely used one is certainly Barry Vercoe's Csound [Bou00], which differs from Pd in being text-based (not GUI-based). This is an advantage in some respects and a disadvantage in others. Csound is better adapted than Pd for batch processing and it handles polyphony much better than Pd does. On the other hand, Pd has a better developed real-time control structure than Csound. Genealogically, Csound belongs to the so-called Music N languages [Mat69, pp.115-172].

Another open-source alternative in wide use is James McCartney's SuperCollider, which is also more text oriented than Pd, but like Pd is explicitly designed for real-time use. SuperCollider has powerful linguistic constructs which make it a more suitable tool than Csound for tasks like writing loops or maintaining complex data structures.

Finally, Pd has a widely-used sibling, Cycling74's commercial program Max/MSP (the others named here are all open source). Both beginners and system managers running multi-user, multi-purpose computer labs will find Max/MSP better supported and documented than Pd. It's possible to take knowledge of Pd and apply it in Max/MSP and vice versa, and even to port patches from one to the other, but the two aren't truly compatible.

Quick introduction to Pd

Pd documents are called *patches*. They correspond roughly to the boxes in the abstract block diagrams shown earlier in this chapter, but in detail they are quite different, because Pd is an implementation environment, not a specification language.

A Pd patch, such as the ones shown in Figure 1.10, consists of a collection of *boxes* connected in a network. The border of a box tells you how its text is interpreted and how the box functions. In part (a) of the figure we see three types of boxes. From top to bottom they are:

- a *message box*. Message boxes, with a flag-shaped border, interpret the text as a message to send whenever the box is activated (by an incoming message or with a pointing device). The message in this case consists simply of the number "21".

- an *object box*. Object boxes have a rectangular border; they interpret the text to create objects when you load a patch. Object boxes may hold hundreds of different classes of objects—including oscillators, envelope generators, and other signal processing modules to be introduced later—depending on the text inside. In this example, the box holds an adder. In most Pd patches, the majority of boxes are of type "object". The first word typed into an object box specifies its *class*, which in this case is just "+". Any additional (blank-space-separated) words appearing in the box are called *creation arguments*, which specify the initial state of the object when it is created.

- a *number box*. Number boxes are a particular type of *GUI box*. Others include push buttons and toggle switches; these will come up later in the examples. The number box has a punched-card-shaped border, with a nick out of its top right corner. Whereas the appearance of an object or message box is fixed when a patch is running, a number box's contents (the text) changes to reflect the current value held by the box. You can also use a number box as a control by clicking and dragging up and down, or by typing values in it.

In Figure 1.10 (part a) the message box, when clicked, sends the message "21" to an object box which adds 13 to it. The lines connecting the boxes carry data from one box to the next; outputs of boxes are on the bottom and inputs on top.

Figure 1.10 (part b) shows a Pd patch which makes a sinusoid with controllable frequency and amplitude. The connecting patch lines are of two types here; the thin ones are for carrying sporadic *messages*, and the thicker ones (connecting the oscillator, the multiplier, and the output dac~

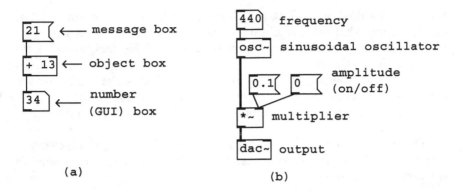

Figure 1.10: (a) three types of boxes in Pd (message, object, and GUI); (b) a simple patch to output a sinusoid.

object) carry digital audio signals. Since Pd is a real-time program, the audio signals flow in a continuous stream. On the other hand, the sporadic messages appear at specific but possibly unpredictable instants in time.

Whether a connection carries messages or signals depends on the box the connection comes from; so, for instance, the + object outputs messages, but the *~ object outputs a signal. The inputs of a given object may or may not accept signals (but they always accept messages, even if only to convert them to signals). As a convention, object boxes with signal inputs or outputs are all named with a trailing tilde ("~") as in "*~" and "osc~".

How to find and run the examples

To run the patches, you must first download, install, and run Pd. Instructions for doing this appear in Pd's on-line HTML documentation, which you can find at http://crca.ucsd.edu/~msp/software.htm.

This book should appear at http:/crca/ucsd/edu/~msp/techniques.htm, possibly in several revisions. Choose the revision that corresponds to the text you're reading (or perhaps just the latest one) and download the archive containing the associated revision of the examples (you may also download an archive of the HTML version of this book for easier access on your machine). The examples should all stay in a single directory, since some of them depend on other files in that directory and might not load them correctly if you have moved things around.

If you do want to copy one of the examples to another directory so that you can build on it (which you're welcome to do), you should either include

the examples directory in Pd's search path (see the Pd documentation) or
else figure out what other files are needed and copy them too. A good way
to find this out is just to run Pd on the relocated file and see what Pd
complains it can't find.

There should be dozens of files in the "examples" folder, including the
examples themselves and the support files. The filenames of the examples
all begin with a letter (A for chapter 1, B for 2, etc.) and a number, as in
"A01.sinewave.pd".

The example patches are also distributed with Pd, but beware that you
may find a different version of the examples which might not correspond to
the text you're reading.

1.9   Examples

Constant amplitude scaler

Example A01.sinewave.pd, shown in Figure 1.11, contains essentially the
simplest possible patch that makes a sound, with only three object boxes.
(There are also comments, and two message boxes to turn Pd's "DSP"
(audio) processing on and off.) The three object boxes are:

| osc ~ |: sinusoidal oscillator. The left hand side input and the output are
digital audio signals. The input is taken to be a (possibly time-varying)
frequency in Hertz. The output is a sinusoid at the specified frequency. If
nothing is connected to the frequency inlet, the creation argument (440 in
this example) is used as the frequency. The output has peak amplitude
one. You may set an initial phase by sending messages (not audio signals)
to the right inlet. The left (frequency) inlet may also be sent messages to
set the frequency, since any inlet that takes an audio signal may also be sent
messages which are automatically converted to the desired audio signal.

| * ~ |: multiplier. This exists in two forms. If a creation argument is
specified (as in this example; it's 0.05), this box multiplies a digital audio
signal (in the left inlet) by the number; messages to the right inlet can
update the number as well. If no argument is given, this box multiplies two
incoming digital audio signals together.

| dac ~ |: audio output device. Depending on your hardware, this might
not actually be a Digital/Analog Converter as the name suggests; but in
general, it allows you to send any audio signal to your computer's audio
output(s). If there are no creation arguments, the default behavior is to
output to channels one and two of the audio hardware; you may specify
alternative channel numbers (one or many) using the creation arguments.
Pd itself may be configured to use two or more output channels, or may not

MAKING A SINE WAVE

Audio computation in Pd is done using "tilde objects" such
as the three below. They use continuous audio streams to
intercommunicate, and also communicate with other
("control") Pd objects using messages.

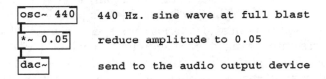

                         440 Hz. sine wave at full blast

                         reduce amplitude to 0.05

                         send to the audio output device

Audio computation can be turned on and off by sending
messages to the global "pd" object as follows:

                                      <-- click these

         ON   OFF

You should see the Pd ("main") window change to reflect
whether audio is on or off. You can also turn audio on and
off using the "audio" menu, but the buttons are provided as
a shortcut.

When DSP is on, you should hear a tone whose pitch is A 440
and whose amplitude is 0.05. If instead you are greeted
with silence, you might want to read the HTML documentation
on setting up audio.

In general when you start a work session with Pd, you will
want to choose "test audio and MIDI" from the help window,
which opens a more comprehensive test patch than this one.

Figure 1.11: The contents of the first Pd example patch: A01.sinewave.pd.

have the audio output device open at all; consult the Pd documentation for details.

The two message boxes show a peculiarity in the way messages are parsed in message boxes. Earlier in Figure 1.10 (part a), the message consisted only of the number 21. When clicked, that box sent the message "21" to its outlet and hence to any objects connected to it. In this current example, the text of the message boxes starts with a semicolon. This is a terminator between messages (so the first message is empty), after which the next word is taken as the name of the recipient of the following message. Thus the message here is "dsp 1" (or "dsp 0") and the message is to be sent, not to any connected objects—there aren't any anyway—but rather, to the object named "pd". This particular object is provided invisibly by the Pd program and you can send it various messages to control Pd's global state, in this case turning audio processing on ("1") and off ("0").

Many more details about the control aspects of Pd, such as the above, are explained in a different series of example patches (the "control examples") in the Pd release, but they will only be touched on here as necessary to demonstrate the audio signal processing techniques that are the subject of this book.

Amplitude control in decibels

Example A02.amplitude.pd shows how to make a crude amplitude control; the active elements are shown in Figure 1.12 (part a). There is one new object class:

[dbtorms] : Decibels to linear amplitude conversion. The "RMS" is a misnomer; it should have been named "dbtoamp", since it really converts from decibels to any linear amplitude unit, be it RMS, peak, or other. An input of 100 dB is normalized to an output of 1. Values greater than 100 are fine (120 will give 10), but values less than or equal to zero will output zero (a zero input would otherwise have output a small positive number). This is a control object, i.e., the numbers going in and out are messages, not signals. (A corresponding object, [dbtorms ∼], is the signal correlate. However, as a signal object this is expensive in CPU time and most often we'll find one way or another to avoid using it.)

The two number boxes are connected to the input and output of the dbtorms object. The input functions as a control; "mouse" on it (click and drag upward or downward) to change the amplitude. It has been set to range from 0 to 80; this is protection for your speakers and ears, and it's wise to build such guardrails into your own patches.

The other number box shows the output of the dbtorms object. It is useless to mouse on this number box, since its outlet is connected nowhere;

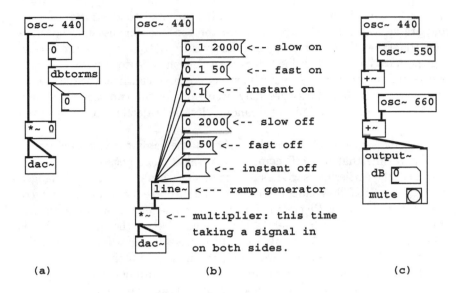

Figure 1.12: The active ingredients to three patches: (a) A02.amplitude.pd; (b) A03.line.pd; (c) A05.output.subpatch.pd.

it is here purely to display its input. Number boxes may be useful as controls, displays, or both, although if you're using it as both there may be some extra work to do.

Smoothed amplitude control with an envelope generator

As Figure 1.6 shows, one way to make smooth amplitude changes in a signal without clicks is to multiply it by the output of an envelope generator as shown in block diagram form in Figure 1.7. This may be implemented in Pd using the `line~` object:

line ~ : envelope generator. The output is a signal which ramps linearly from one value to another over time, as determined by the messages received. The inlets take messages to specify target values (left inlet) and time delays (right inlet). Because of a general rule of Pd messages, a pair of numbers sent to the left inlet suffices to specify a target value and a time together. The time is in milliseconds (taking into account the sample rate), and the target value is unitless, or in other words, its output range should conform to whatever input it may be connected to.

Example A03.line.pd demonstrates the use of a `line~` object to control the amplitude of a sinusoid. The active part is shown in Figure 1.12

(part b). The six message boxes are all connected to the line~ object, and are activated by clicking on them; the top one, for instance, specifies that the line~ ramp (starting at wherever its output was before receiving the message) to the value 0.1 over two seconds. After the two seconds elapse, unless other messages have arrived in the meantime, the output remains steady at 0.1. Messages may arrive before the two seconds elapse, in which case the line~ object abandons its old trajectory and takes up a new one.

Two messages to line~ might arrive at the same time or so close together in time that no DSP computation takes place between the two; in this case, the earlier message has no effect, since line~ won't have changed its output yet to follow the first message, and its current output, unchanged, is then used as a starting point for the second segment. An exception to this rule is that, if line~ gets a time value of zero, the output value is immediately set to the new value and further segments will start from the new value; thus, by sending two pairs, the first with a time value of zero and the second with a nonzero time value, one can independently specify the beginning and end values of a segment in line~'s output.

The treatment of line~'s right inlet is unusual among Pd objects in that it forgets old values; a message with a single number such as "0.1" is always equivalent to the pair, "0.1 0". Almost any other object will retain the previous value for the right inlet, instead of resetting it to zero.

Example A04.line2.pd   shows the line~ object's output graphically. Using the various message boxes, you can recreate the effects shown in Figure 1.6.

Major triad

Example A05.output.subpatch.pd, whose active ingredients are shown in Figure 1.12 (part c), presents three sinusoids with frequencies in the ratio 4:5:6, so that the lower two are separated by a major third, the upper two by a minor third, and the top and bottom by a fifth. The lowest frequency is 440, equal to A above middle C, or MIDI 69. The others are approximately four and seven half-steps higher, respectively. The three have equal amplitudes.

The amplitude control in this example is taken care of by a new object called output~. This isn't a built-in object of Pd, but is itself a Pd patch which lives in a file, "output.pd". (You can see the internals of output~ by opening the properties menu for the box and selecting "open".) You get two controls, one for amplitude in dB (100 meaning "unit gain"), and a "mute" button. Pd's audio processing is turned on automatically when you set the output level—this might not be the best behavior in general, but it's appropriate for these example patches. The mechanism for embedding one Pd patch as an object box inside another is discussed in Section 4.7.

Conversion between frequency and pitch

Example A06.frequency.pd (Figure 1.13) shows Pd's object for converting pitch to frequency units (mtof, meaning "MIDI to frequency") and its inverse ftom. We also introduce two other object classes, send and receive.

mtof , ftom : convert MIDI pitch to frequency units according to the Pitch/Frequency Conversion Formulas (Page 7). Inputs and outputs are messages ("tilde" equivalents of the two also exist, although like dbtorms~ they're expensive in CPU time). The ftom object's output is -1500 if the input is zero or negative; and likewise, if you give mtof -1500 or lower it outputs zero.

receive , r : Receive messages non-locally. The receive object, which may be abbreviated as "r", waits for non-local messages to be sent by a send object (described below) or by a message box using redirection (the ";" feature discussed in the earlier example, A01.sinewave.pd). The argument (such as "frequency" and "pitch" in this example) is the name to which messages are sent. Multiple receive objects may share the same name, in which case any message sent to that name will go to all of them.

send , s : The send object, which may be abbreviated as "s", directs messages to receive objects.

Two new properties of number boxes are used here. Earlier we've used them as controls or as displays; here, the two number boxes each function as both. If a number box gets a number in its inlet, it not only displays the number but also repeats the number to its output. However, a number box

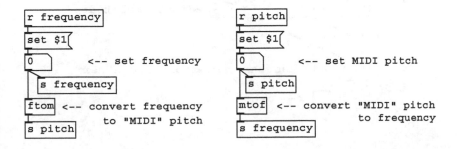

Figure 1.13: Conversion between pitch and frequency in A06.frequency.pd.

may also be sent a "set" message, such as "set 55" for example. This would set the value of the number box to 55 (and display it) but not cause the output that would result from the simple "55" message. In this case, numbers coming from the two **receive** objects are formatted (using message boxes) to read "set 55" instead of just "55", and so on. (The special word "$1" is replaced by the incoming number.) This is done because otherwise we would have an infinite loop: frequency would change pitch which would change frequency and so on forever, or at least until something broke.

More additive synthesis

The major triad (Example A06.frequency.pd, Page 22) shows one way to combine several sinusoids together by summing. There are many other possible ways to organize collections of sinusoids, of which we'll show two. Example A07.fusion.pd (Figure 1.14) shows four oscillators, whose frequencies are tuned in the ratio 1:2:3:4, with relative amplitudes 1, 0.1, 0.2, and 0.5. The amplitudes are set by multiplying the outputs of the oscillators (the *~ objects below the oscillators).

The second, third, and fourth oscillators are turned on and off using a *toggle switch*. This is a graphical control, like the number box introduced earlier. The toggle switch puts out 1 and 0 alternately when clicked

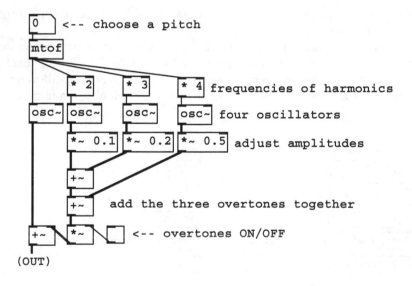

Figure 1.14: Additive synthesis using harmonically tuned oscillators.

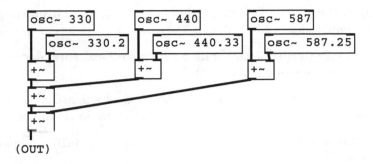

Figure 1.15: Additive synthesis: six oscillators arranged into three beating pairs.

on with the mouse. This value is multiplied by the sum of the second, third, and fourth oscillators, effectively turning them on and off.

Even when all four oscillators are combined (with the toggle switch in the "1" position), the result fuses into a single tone, heard at the pitch of the leftmost oscillator. In effect this patch sums a four-term Fourier series to generate a complex, periodic waveform.

Example A08.beating.pd (Figure 1.15) shows another possibility, in which six oscillators are tuned into three pairs of neighbors, for instance 330 and 330.2 Hertz. These pairs slip into and out of phase with each other, so that the amplitude of the sum changes over time. Called *beating*, this phenomenon is frequently used for musical effects.

Oscillators may be combined in other ways besides simply summing their output, and a wide range of resulting sounds is available. Example A09.frequency.mod.pd (not shown here) demonstrates *frequency modulation* synthesis, in which one oscillator controls another's frequency. This will be more fully described in Chapter 5.

Exercises

1. A sinusoid (Page 1) has initial phase $\phi = 0$ and angular frequency $\omega = \pi/10$. What is its period in samples? What is the phase at sample number $n = 10$?

2. Two sinusoids have periods of 20 and 30 samples, respectively. What is the period of the sum of the two?

3. If 0 dB corresponds to an amplitude of 1, how many dB corresponds to amplitudes of 1.5, 2, 3, and 5?

4. Two uncorrelated signals of RMS amplitude 3 and 4 are added; what's the RMS amplitude of the sum?

5. How many uncorrelated signals, all of equal amplitude, would you have to add to get a signal that is 9 dB greater in amplitude?

6. What is the angular frequency of middle C at 44100 samples per second?

7. Two sinusoids play at middle C (MIDI 60) and the neighboring C sharp (MIDI 61). What is the difference, in Hertz, between their frequencies?

8. How many cents is the interval between the seventh and the eighth harmonic of a periodic signal?

9. If an audio signal $x[n], n = 0, ..., N - 1$ has peak amplitude 1, what is the minimum possible RMS amplitude? What is the maximum possible?

Chapter 2

Wavetables and Samplers

In Chapter 1 we treated audio signals as if they always flowed by in a continuous stream at some sample rate. The sample rate isn't really a quality of the audio signal, but rather it specifies how fast the individual samples should flow into or out of the computer. But audio signals are at bottom just sequences of numbers, and in practice there is no requirement that they be "played" sequentially. Another, complementary view is that they can be stored in memory, and, later, they can be read back in any order—forward, backward, back and forth, or totally at random. An inexhaustible range of new possibilities opens up.

For many years (roughly 1950-1990), magnetic tape served as the main storage medium for sounds. Tapes were passed back and forth across magnetic pickups to play the signals back in real time. Since 1995 or so, the predominant way of storing sounds has been to keep them as digital audio signals, which are read back with much greater freedom and facility than were the magnetic tapes. Many modes of use dating from the tape era are still current, including cutting, duplication, speed change, and time reversal. Other techniques, such as *waveshaping*, have come into their own only in the digital era.

Suppose we have a stored digital audio signal, which is just a sequence of samples (i.e., numbers) $x[n]$ for $n = 0, ..., N-1$, where N is the length of the sequence. Then if we have an input signal $y[n]$ (which we can imagine to be flowing in real time), we can use its values as indices to look up values of the stored signal $x[n]$. This operation, called *wavetable lookup*, gives us a new signal, $z[n]$, calculated as:

$$z[n] = x[y[n]]$$

Schematically we represent this operation as shown in Figure 2.1.

IN

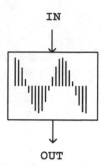

OUT

Figure 2.1: Diagram for wavetable lookup. The input is in samples, ranging approximately from 0 to the wavetable's size N, depending on the interpolation scheme.

Two complications arise. First, the input values, $y[n]$, might lie outside the range $0, ..., N - 1$, in which case the wavetable $x[n]$ has no value and the expression for the output $z[n]$ is undefined. In this situation we might choose to *clip* the input, that is, to substitute 0 for anything negative and $N - 1$ for anything N or greater. Alternatively, we might prefer to wrap the input around end to end. Here we'll adopt the convention that out-of-range samples are always clipped; when we need wraparound, we'll introduce another signal processing operation to do it for us.

The second complication is that the input values need not be integers; in other words they might fall between the points of the wavetable. In general, this is addressed by choosing some scheme for interpolating between the points of the wavetable. For the moment, though, we'll just round down to the nearest integer below the input. This is called *non-interpolating* wavetable lookup, and its full definition is:

$$z[n] = \begin{cases} x[\lfloor y[n] \rfloor] & \text{if } 0 \le y[n] < N - 1 \\ x[0] & \text{if } y[n] < 0 \\ x[N - 1] & \text{if } y[n] \ge N - 1 \end{cases}$$

(where $\lfloor y[n] \rfloor$ means, "the greatest integer not exceeding $y[n]$").

Pictorally, we use $y[0]$ (a number) as a location on the horizontal axis of the wavetable shown in Figure 2.1, and the output, $z[0]$, is whatever we get on the vertical axis; and the same for $y[1]$ and $z[1]$ and so on. The "natural" range for the input $y[n]$ is $0 \le y[n] < N$. This is different from the usual range of an audio signal suitable for output from the computer, which ranges from -1 to 1 in our units. We'll see later that the usable range

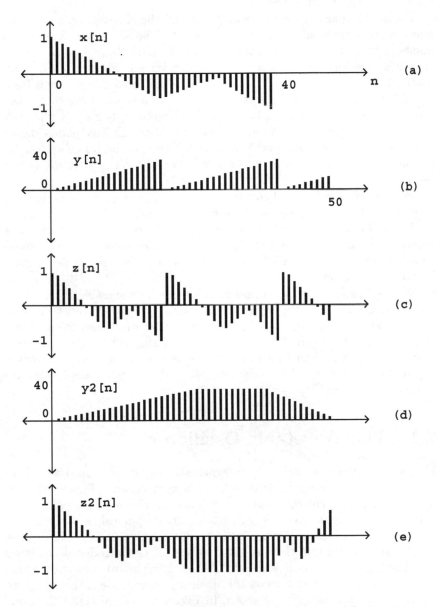

Figure 2.2: Wavetable lookup: (a) a wavetable; (b) and (d) signal inputs for lookup; (c) and (e) the corresponding outputs.

of input values, from 0 to N for non-interpolating lookup, shrinks slightly if interpolating lookup is used.

Figure 2.2 (part a) shows a wavetable and the result of using two different input signals as lookup indices into it. The wavetable contains 40 points, which are numbered from 0 to 39. In part (b), a *sawtooth wave* is used as the input signal $y[n]$. A sawtooth wave is nothing but a ramp function repeated end to end. In this example the sawtooth's range is from 0 to 40 (this is shown in the vertical axis). The sawtooth wave thus scans the wavetable from left to right—from the beginning point 0 to the endpoint 39—and does so every time it repeats. Over the fifty points shown in Figure 2.2 (part b) the sawtooth wave makes two and a half cycles. Its period is twenty samples, or in other words the frequency (in cycles per second) is $R/20$.

Part (c) of Figure 2.2 shows the result of applying wavetable lookup, using the table $x[n]$, to the signal $y[n]$. Since the sawtooth input simply reads out the contents of the wavetable from left to right, repeatedly, at a constant rate of precession, the result will be a new periodic signal, whose waveform (shape) is derived from $x[n]$ and whose frequency is determined by the sawtooth wave $y[n]$.

Parts (d) and (e) show an example where the wavetable is read in a nonuniform way; since the input signal rises from 0 to N and then later recedes to 0, we see the wavetable appear first forward, then frozen at its endpoint, then backward. The table is scanned from left to right and then, more quickly, from right to left. As in the previous example the incoming signal controls the speed of precession while the output's amplitudes are those of the wavetable.

2.1   The Wavetable Oscillator

Figure 2.2 suggests an easy way to synthesize any desired fixed waveform at any desired frequency, using the block diagram shown in Figure 2.3. The upper block is an oscillator—not the sinusoidal oscillator we saw earlier, but one that produces sawtooth waves instead. Its output values, as indicated at the left of the block, should range from 0 to the wavetable size N. This is used as an index into the wavetable lookup block (introduced in Figure 2.1), resulting in a periodic waveform. Figure 2.3 (part b) adds an envelope generator and a multiplier to control the output amplitude in the same way as for the sinusoidal oscillator shown in Figure 1.7 (Page 11). Often, one uses a wavetable with (RMS or peak) amplitude 1, so that the amplitude of the output is just the magnitude of the envelope generator's output.

Wavetable oscillators are often used to synthesize sounds with specified, static spectra. To do this, you can pre-compute N samples of any wave-

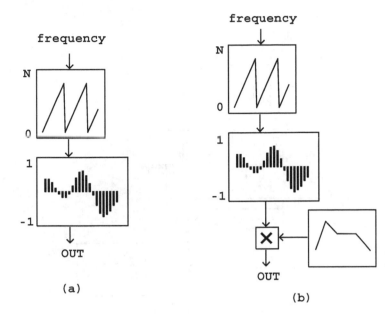

Figure 2.3: Block diagrams: (a) for a wavetable lookup oscillator; (b) with amplitude control by an envelope generator.

form of period N (angular frequency $2\pi/N$) by adding up the elements of the Fourier Series (Page 12). The computation involved in setting up the wavetable at first might be significant, but this may be done in advance of the synthesis process, which might take place in real time.

While direct additive synthesis of complex waveforms, as shown in Chapter 1, is in principle infinitely flexible as a technique for producing time-varying timbres, wavetable synthesis is much less expensive in terms of computation but requires switching wavetables to change the timbre. An intermediate technique, more flexible and expensive than simple wavetable synthesis but less flexible and less expensive than additive synthesis, is to create time-varying mixtures between a small number of fixed wavetables. If the number of wavetables is only two, this is in effect a cross-fade between the two waveforms, as diagrammed in Figure 2.4. Suppose we wish to use some signal $0 \leq x[n] \leq 1$ to control the relative strengths of the two waveforms, so that, if $x[n] = 0$, we get the first one and if $x[n] = 1$ we get the second. Denoting the two signals to be cross-faded by $y[n]$ and $z[n]$, we compute the signal

$$(1 - x[n])y[n] + x[n]z[n]$$

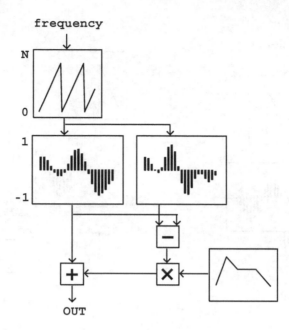

Figure 2.4: Block diagram for cross-fading between two wavetables.

or, equivalently and usually more efficient to calculate,

$$y[n] + x[n](z[n] - y[n])$$

This computation is diagrammed in Figure 2.4.

When using this technique to cross-fade between wavetable oscillators, it might be desirable to keep the phases of corresponding partials the same across the wavetables, so that their amplitudes combine additively when they are mixed. On the other hand, if arbitrary wavetables are used (borrowed, for instance, from a recorded sound) there will be a phasing effect as the different waveforms are mixed.

This scheme can be extended in a daisy chain to move along a continuous path between a succession of timbres. Alternatively, or in combination with daisy-chaining, cross-fading may be used to interpolate between two different timbres, for example as a function of musical dynamic. To do this you would prepare two or even several waveforms of a single synthetic voice played at different dynamics, and interpolate between successive ones as a function of the output dynamic you want.

2.2 Sampling

"Sampling" is nothing more than recording a live signal into a wavetable, and then later playing it out again. (In commercial samplers the entire wavetable is usually called a "sample" but to avoid confusion we'll only use the word "sample" here to mean a single number in an audio signal.)

At its simplest, a sampler is simply a wavetable oscillator, as was shown in Figure 2.3. However, in the earlier discussion we imagined playing the oscillator back at a frequency high enough to be perceived as a pitch, at least 30 Hertz or so. In the case of sampling, the frequency is usually lower than 30 Hertz, and so the period, at least 1/30 second and perhaps much more, is long enough that you can hear the individual cycles as separate events.

Going back to Figure 2.2, suppose that instead of 40 points the wavetable $x[n]$ is a one-second recording, at an original sample rate of 44100, so that it has 44100 points; and let $y[n]$ in part (b) of the figure have a period of 22050 samples. This corresponds to a frequency of 2 Hertz. But what we hear is not a pitched sound at 2 cycles per second (that's too slow to hear as a pitch) but rather, we hear the original recording $x[n]$ played back repeatedly at double speed. We've just reinvented the sampler.

In general, if we assume the sample rate R of the recording is the same as the output sample rate, if the wavetable has N samples, and if we index it with a sawtooth wave of period M, the sample is sped up or slowed down by a factor of N/M, equal to Nf/R if f is the frequency in Hertz of the sawtooth. If we denote the transposition factor by t (so that, for instance, $t = 3/2$ means transposing upward a perfect fifth), and if we denote the transposition in half-steps by h, then we get the Transposition Formulas for Looping Wavetables:

$$t = N/M = Nf/R$$

$$h = 12 \log_2 \left(\frac{N}{M} \right) = 12 \log_2 \left(\frac{Nf}{R} \right)$$

Frequently the desired transposition in half-steps (h) is known and the formula must be solved for either f or N:

$$f = \frac{2^{h/12} R}{N}$$

$$N = \frac{2^{h/12} R}{f}$$

So far we have used a sawtooth as the input wave $y[t]$, but, as suggested in parts (d) and (e) of Figure 2.2, we could use anything we like as an

input signal. In general, the transposition may be time dependent and is
controlled by the rate of change of the input signal.

The transposition multiple t and the transposition in half-steps h are
then given by the Momentary Transposition Formulas for Wavetables:

$$t[n] = |y[n] - y[n-1]|$$

$$h[n] = 12\log_2 |y[n] - y[n-1]|$$

(Here the enclosing bars "|" mean absolute value.) For example, if $y[n] = n$,
then $z[n] = x[n]$ so we hear the wavetable at its original pitch, and this is
what the formula predicts since, in that case,

$$y[n] - y[n-1] = 1$$

On the other hand, if $y[n] = 2n$, then the wavetable is transposed up an
octave, consistent with

$$y[n] - y[n-1] = 2$$

If values of $y[n]$ are decreasing with n, you hear the sample backward, but
the transposition formula still gives a positive multiplier. This all agrees
with the earlier Transposition Formula for Looping Wavetables; if a sawtooth ranges from 0 to N, f times per second, the difference of successive
samples is just Nf/R—except at the sample at the beginning of each new
cycle.

It's well known that transposing a recording also transposes its timbre—
this is the "chipmunk" effect. Not only are any periodicities (such as might
give rise to pitch) transposed, but so are the frequencies of the overtones.
Some timbres, notably those of vocal sounds, have characteristic frequency
ranges in which overtones are stronger than other nearby ones. Such frequency ranges are also transposed, and this is is heard as a timbre change.
In language that will be made more precise in Section 5.1, we say that the
spectral envelope is transposed along with the pitch or pitches.

In both this and the preceding section, we have considered playing
wavetables periodically. In Section 2.1 the playback repeated quickly enough
that the repetition gives rise to a pitch, say between 30 and 4000 times per
second, roughly the range of a piano. In the current section we assumed
a wavetable one second long, and in this case "reasonable" transposition
factors (less than four octaves up) would give rise to a rate of repetition
below 30, usually much lower, and going down as low as we wish.

The number 30 is significant for another reason: it is roughly the maximum number of separate events the ear can discern per second; for instance,
30 vocal phonemes, or melodic notes, or attacks of a snare drum are about
the most we can hope to crowd into a second before our ability to distinguish them breaks down.

A continuum exists between samplers and wavetable oscillators, in that the patch of Figure 2.3 can either be regarded as a sampler (if the frequency of repetition is less than about 20 Hertz) or as a wavetable oscillator (if the frequency is greater than about 40 Hertz). It is possible to move continuously between the two regimes. Furthermore, it is not necessary to play an entire wavetable in a loop; with a bit more arithmetic we can choose sub-segments of the wavetable, and these can change in length and location continuously as the wavetable is played.

The practice of playing many small segments of a wavetable in rapid succession is often called *granular synthesis*. For much more discussion of the possibilities, see [Roa01].

Figure 2.5 shows how to build a very simple looping sampler. In the figure, if the frequency is f and the segment size in samples is s, the output transposition factor is given by $t = fs/R$, where R is the sample rate at which the wavetable was recorded (which need not equal the sample rate the block diagram is working at.) In practice, this equation must usually be solved for either f or s to attain a desired transposition.

In the figure, a sawtooth oscillator controls the location of wavetable lookup, but the lower and upper values of the sawtooth aren't statically specified as they were in Figure 2.3; rather, the sawtooth oscillator simply ranges from 0 to 1 in value and the range is adjusted to select a desired segment of samples in the wavetable.

It might be desirable to specify the segment's location l either as its left-hand edge (its lower bound) or else as the segment's midpoint; in either case we specify the length s as a separate parameter. In the first case, we start by multiplying the sawtooth by s, so that it then ranges from 0 to s; then we add l so that it now ranges from l to $l + s$. In order to specify the location as the segment's midpoint, we first subtract $1/2$ from the sawtooth (so that it ranges from $-1/2$ to $1/2$), and then as before multiply by s (so that it now ranges from $-s/2$ to $s/2$) and add l to give a range from $l - s/2$ to $l + s/2$.

In the looping sampler, we will need to worry about maintaining continuity between the beginning and the end of segments of the wavetable; we'll take this up in the next section.

A further detail is that, if the segment size and location are changing with time (they might be digital audio signals themselves, for instance), they will affect the transposition factor, and the pitch or timbre of the output signal might waver up and down as a result. The simplest way to avoid this problem is to synchronize changes in the values of s and l with the regular discontinuities of the sawtooth; since the signal jumps discontinuously there, the transposition is not really defined there anyway, and, if you are enveloping to hide the discontinuity, the effects of changes in s and l are hidden as well.

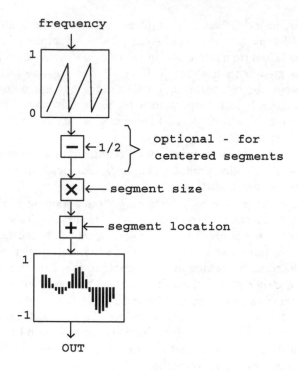

Figure 2.5: A simple looping sampler, as yet with no amplitude control. There are inputs to control the frequency and the segment size and location. The "-" operation is included if we wish the segment location to be specified as the segment's midpoint; otherwise we specify the location of the left end of the segment.

2.3 Enveloping Samplers

In the previous section we considered reading a wavetable either sporadically or repeatedly to make a sampler. In most real applications we must deal with getting the samples to start and stop cleanly, so that the output signal doesn't jump discontinuously at the beginnings and ends of samples. This discontinuity can sound like a click or a thump depending on the wavetable.

The easiest way to do this, assuming we will always play a wavetable completely from beginning to end, is simply to prepare it in advance so that it fades in cleanly at the beginning and out cleanly at the end. This may even be done when the wavetable is sampled live, by multiplying the

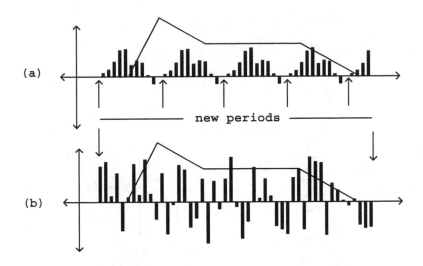

Figure 2.6: Differing envelope requirements for oscillators and samplers: (a) in an oscillator, the envelope can be chosen to conform to any desired timescale; (b) when the wavetable is a recorded sound, it's up to you to get the envelope to zero before you hit the end of the wavetable for the first time.

input signal by a line segment envelope timed to match the length of the recording.

In many situations, however, it is either inconvenient or impossible to pre-envelope the wavetable—for example, we might want to play only part of it back, or we may want to change the sharpness of the enveloping dynamically. In Section 1.5 we had already seen how to control the amplitude of sinusoidal oscillators using multiplication by a ramp function (also known as an envelope generator), and we built this notion into the wavetable oscillators of Figures 2.3 and 2.4. This also works fine for turning samplers on and off to avoid discontinuities, but with one major difference: whereas in wavetable synthesis, we were free to assume that the waveform lines up end to end, so that we may choose any envelope timing we want, in the case of sampling using unprepared waveforms, we are obliged to get the envelope generator's output to zero by the time we reach the end of the wavetable for the first time. This situation is pictured in Figure 2.6.

In situations where an arbitrary wavetable must be repeated as needed, the simplest way to make the looping work continuously is to arrange for amplitude change to be synchronized with the looping, using a separate

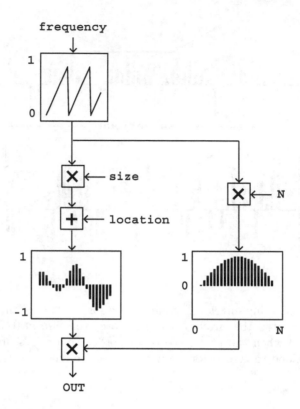

Figure 2.7: A sampler as in Figure 2.5, but with an additional wavetable lookup for enveloping.

wavetable (the envelope). This may be implemented as shown in Figure 2.7. A single sawtooth oscillator is used to calculate lookup indices for two wavetables, one holding the recorded sound, and the other, an envelope shape. The main thing to worry about is getting the inputs of the two wavetables each into its own appropriate range.

In many situations it is desirable to combine two or more copies of the looping wavetable sampler at the same frequency and at a specified phase relationship. This may be done so that when any particular one is at the end of its segment, one or more others is in the middle of the same segment, so that the aggregate is continuously making sound. To accomplish this, we need a way to generate two or more sawtooth waves at the desired phase relationship that we can use in place of the oscillator at the top of Figure 2.7. We can start with a single sawtooth wave and then produce others at

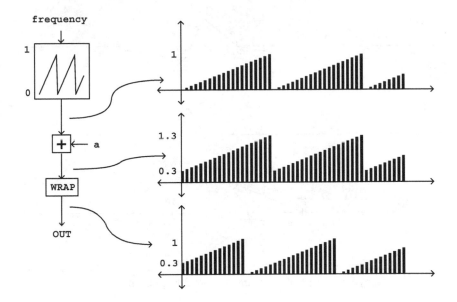

Figure 2.8: A technique for generating two or more sawtooth waves with fixed phase relationships between them. The relative phase is controlled by the parameter a (which takes the value 0.3 in the graphed signals). The "wrap" operation computes the fractional part of its input.

fixed phase relationships with the first one. If we wish a sawtooth which is, say, a cycles ahead of the first one, we simply add the parameter a and then take the fractional part, which is the desired new sawtooth wave, as shown in Figure 2.8.

2.4 Timbre Stretching

The wavetable oscillator of Section 2.1, which we extended in Section 2.2 to encompass grabbing waveforms from arbitrary wavetables such as recorded sounds, may additionally be extended in a complementary way, that we'll refer to as *timbre stretching*, for reasons we'll develop in this section. There are also many other possible ways to extend wavetable synthesis, using for instance frequency modulation and waveshaping, but we'll leave them to later chapters.

The central idea of timbre stretching is to reconsider the idea of the wavetable oscillator as a mechanism for playing a stored wavetable (or part of one) end to end. There is no reason the end of one cycle has to co-

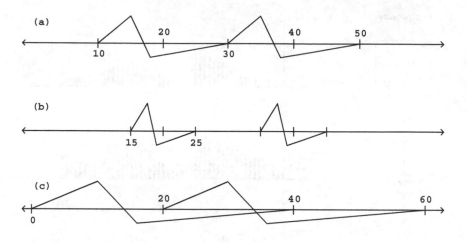

Figure 2.9: A waveform is played at a period of 20 samples: (a) at 100 percent duty cycle; (b) at 50 percent; (c) at 200 percent

incide with the beginning of another. Instead, we could ask for copies of the waveform to be spaced with alternating segments of silence; or, going in the opposite direction, the waveform copies could be spaced more closely together so that they overlap. The single parameter available in Section 2.1—the frequency—has been heretofore used to control two separate aspects of the output: the period at which we start new copies of the waveform, and also the length of each individual copy. The idea of timbre stretching is to control the two independently.

Figure 2.9 shows the result of playing a wavetable in three ways. In each case the output waveform has period 20; in other words, the output frequency is $R/20$ if R is the output sample rate. In part (a) of the figure, each copy of the waveform is played over 20 samples, so that the wave form fits exactly into the cycle with no gaps and no overlap. In part (b), although the period is still 20, the waveform is compressed into the middle half of the period (10 samples); or in other words, the *duty cycle*—the relative amount of time the waveform fills the cycle—equals 50 percent. The remaining 50 percent of the time, the output is zero.

In part (c), the waveform is stretched to 40 samples, and since it is still repeated every 20 samples, the waveforms overlap two to one. The duty cycle is thus 200 percent.

Suppose now that the 100 percent duty cycle waveform has a Fourier series (Section 1.7) equal to:

$$x_{100}[n] = a_0 + a_1 \cos\left(\omega n + \phi_1\right) + a_2 \cos\left(2\omega n + \phi_2\right) + \cdots$$

where ω is the angular frequency (equal to $\pi/10$ in our example since the period is 20.) To simplify this example we won't worry about where the series must end, and will just let it go on forever.

We would like to relate this to the Fourier series of the other two waveforms in the example, in order to show how changing the duty cycle changes the timbre of the result. For the 50 percent duty cycle case (calling the signal $x_{50}[n]$), we observe that the waveform, if we replicate it out of phase by a half period and add the two, gives exactly the original waveform at twice the frequency:

$$x_{100}[2n] = x_{50}[n] + x_{50}[n + \frac{\pi}{\omega}]$$

where ω is the angular frequency (and so π/ω is half the period) of both signals. So if we denote the Fourier series of $x_{50}[n]$ as:

$$x_{50}[n] = b_0 + b_1 \cos(\omega n + \theta_1) + b_2 \cos(2\omega n + \theta_2) + \cdots$$

and substitute the Fourier series for all three terms above, we get:

$$a_0 + a_1 \cos(2\omega n + \phi_1) + a_2 \cos(4\omega n + \phi_2) + \cdots$$

$$= b_0 + b_1 \cos(\omega n + \theta_1) + b_2 \cos(2\omega n + \theta_2) + \cdots$$

$$+ b_0 + b_1 \cos(\omega n + \pi + \theta_1) + b_2 \cos(2\omega n + 2\pi + \theta_2) + \cdots$$

$$= 2b_0 + 2b_2 \cos(2\omega n + \theta_2) + 2b_4 \cos(4\omega n + \theta_4) + \cdots$$

and so

$$a_0 = 2b_0, \quad a_1 = 2b_2, \quad a_2 = 2b_4$$

and so on: the even partials of x_{50}, at least, are obtained by stretching the partials of x_{100} out twice as far. (We don't yet know about the odd partials of x_{50}, and these might be in line with the even ones or not, depending on factors we can't control yet. Suffice it to say for the moment, that if the waveform connects smoothly with the horizontal axis at both ends, the odd partials will act globally like the even ones. To make this more exact we'll need Fourier analysis, which is developed in Chapter 9.)

Similarly, x_{100} and x_{200} are related in exactly the same way:

$$x_{200}[2n] = x_{100}[n] + x_{100}[n + \frac{\pi}{\omega}]$$

so that, if the amplitudes of the fourier series of x_{200} are denoted by c_0, c_1, ..., we get:

$$c_0 = 2a_0, c_1 = 2a_2, c_2 = 2a_4, \ldots$$

so that the partials of x_{200} are those of x_{100} shrunk, by half, to the left.

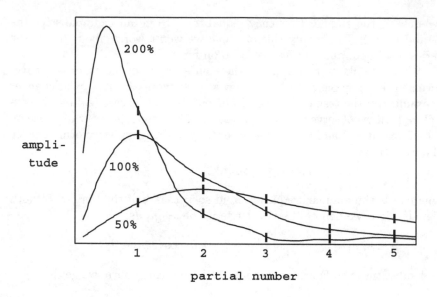

Figure 2.10: The Fourier series magnitudes for the waveforms shown in Figure 2.9. The horizontal axis is the harmonic number. We only "hear" the coefficients for integer harmonic numbers; the continuous curves are the "ideal" contours.

We see that squeezing the waveform by a factor of 2 has the effect of stretching the Fourier series out by two, and on the other hand stretching the waveform by a factor of two squeezes the Fourier series by two. By the same sort of argument, in general it turns out that stretching the waveform by a factor of any positive number f squeezes the overtones, in frequency, by the reciprocal $1/f$—at least approximately, and the approximation is at least fairly good if the waveform "behaves well" at its ends. (As we'll see later, the waveform can always be forced to behave at least reasonably well by enveloping it as in Figure 2.7.)

Figure 2.10 shows the spectra of the three waveforms—or in other words the one waveform at three duty cycles—of Figure 2.9. The figure emphasizes the relationship between the three spectra by drawing curves through each, which, on inspection, turn out to be the same curve, only stretched differently; as the duty cycle goes up, the curve is both compressed to the left (the frequencies all drop) and amplified (stretched upward).

The continuous curves have a very simple interpretation. Imagine squeezing the waveform into some tiny duty cycle, say 1 percent. The contour will be stretched by a factor of 100. Working backward, this would allow us to interpolate between each pair of consecutive points of the 100 percent duty cycle contour (the original one) with 99 new ones. Already in the figure the 50 percent duty cycle trace defines the curve with twice the resolution of the original one. In the limit, as the duty cycle gets arbitrarily small, the spectrum is filled in more and more densely; and the limit is the "true" spectrum of the waveform.

This "true" spectrum is only audible at suitably low duty cycles, though. The 200 percent duty cycle example actually misses the peak in the ideal (continuous) spectrum because the peak falls below the first harmonic. In general, higher duty cycles sample the ideal curve at lower resolutions.

Timbre stretching is an extremely powerful technique for generating sounds with systematically variable spectra. Combined with the possibilities of mixtures of waveforms (Section 2.1) and of snatching endlessly variable waveforms from recorded samples (Section 2.2), it is possible to generate all sorts of sounds. For example, the block diagram of Figure 2.7 gives us a way to to grab and stretch timbres from a recorded wavetable. When the "frequency" parameter f is high enough to be audible as a pitch, the "size" parameter s can be thought of as controlling timbre stretch, via the formula $t = fs/R$ from Section 2.2, where we now reinterpret t as the factor by which the timbre is to be stretched.

2.5 Interpolation

As mentioned before, interpolation schemes are often used to increase the accuracy of table lookup. Here we will give a somewhat simplified account of the effects of table sizes and interpolation schemes on the result of table lookup.

To speak of error in table lookup, we must view the wavetable as a sampled version of an underlying function. When we ask for a value of the underlying function which lies between the points of the wavetable, the error is the difference between the result of the wavetable lookup and the "ideal" value of the function at that point. The most revealing study of wavetable lookup error assumes that the underlying function is a sinusoid (Page 1). We can then understand what happens to other wavetables by considering them as superpositions (sums) of sinusoids.

The accuracy of lookup from a wavetable containing a sinusoid depends on two factors: the quality of the interpolation scheme, and the period of the sinusoid. In general, the longer the period of the sinusoid, the more accurate the result.

In the case of a synthetic wavetable, we might know its sinusoidal components from having specified them—in which case the issue becomes one of choosing a wavetable size appropriately, when calculating the wavetable, to match the interpolation algorithm and meet the desired standard of accuracy. In the case of recorded sounds, the accuracy analysis might lead us to adjust the sample rate of the recording, either at the outset or else by resampling later.

Interpolation error for a sinusoidal wavetable can have two components: first, the continuous signal (the theoretical result of reading the wavetable continuously in time, as if the output sample rate were infinite) might not be a pure sinusoid; and second, the amplitude might be wrong. (It is possible to get phase errors as well, but only through carelessness.)

In this treatment we'll only consider polynomial interpolation schemes such as rounding, linear interpolation, and cubic interpolation. These schemes amount to evaluating polynomials (of degree zero, one, and three, respectively) in the interstices between points of the wavetable. The idea is that, for any index x, we choose a nearby reference point x_0, and let the output be calculated by some polynomial:

$$y_{\mathrm{INT}}(x) = a_0 + a_1(x - x_0) + a_2(x - x_0)^2 + \cdots + a_n(x - x_0)^n$$

Usually we choose the polynomial which passes through the $n + 1$ nearest points of the wavetable. For 1-point interpolation (a zero-degree polynomial) this means letting a_0 equal the nearest point of the wavetable. For two-point interpolation, we draw a line segment between the two points of the wavetable on either side of the desired point x. We can let x_0 be the closest integer to the left of x (which we write as $\lfloor x \rfloor$) and then the formula for linear interpolation is:

$$y_{\mathrm{INT}}(x) = y[x_0] + (y[x_0 + 1] - y[x_0]) \cdot (x - x_0)$$

which is a polynomial, as in the previous formula, with

$$a_0 = y[x_0]$$

$$a_1 = y[x_0 + 1] - y[x_0]$$

In general, you can fit exactly one polynomial of degree $n - 1$ through any n points as long as their x values are all different.

Figure 2.11 shows the effect of using linear (two-point) interpolation to fill in a sinusoid of period 6. At the top are three traces: the original sinusoid, the linearly-interpolated result of using 6 points per period to represent the sinusoid, and finally, another sinusoid, of slightly smaller amplitude, which better matches the six-segment waveform. The error introduced by replacing the original sinusoid by the linearly interpolated version has two components: first, a (barely perceptible) change in amplitude, and second, a (very perceptible) distortion of the wave shape.

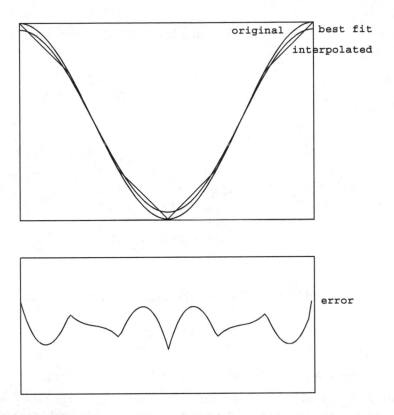

original   best fit

interpolated

error

Figure 2.11: Linear interpolation of a sinusoid: (upper graph) the original sinusoid, the interpolated sinusoid, and the best sinusoidal fit back to the interpolated version; (lower graph) the error, rescaled vertically.

The bottom graph in the figure shows the difference between the interpolated waveform and the best-fitting sinusoid. This is a residual signal all of whose energy lies in overtones of the original sinusoid. As the number of points increases, the error decreases in magnitude. Since the error is the difference between a sinusoid and a sequence of approximating line segments, the magnitude of the error is roughly proportional to the square of the phase difference between each pair of points, or in other words, inversely proportional to the square of the number of points in the wavetable. Put another way, wavetable error decreases by 12 dB each time the table doubles in size. (This rule of thumb is only good for tables with 4 or more points.)

Table 2.1: RMS error for table lookup using 1, 2, and 4 point interpolation at various table sizes.

period	interpolation points		
	1	2	4
2	-1.2	-17.1	-20.2
3	-2.0	-11.9	-15.5
4	-4.2	-17.1	-24.8
8	-10.0	-29.6	-48.4
16	-15.9	-41.8	-72.5
32	-21.9	-53.8	-96.5
64	-27.9	-65.9	-120.6
128	-34.0	-77.9	-144.7

Four-point (cubic) interpolation works similarly. The interpolation formula is:

$$y_{\mathrm{INT}}(x) =$$

$$-f(f-1)(f-2)/6 \cdot y[x_0 - 1] + (f+1)(f-1)(f-2)/2 \cdot y[x_0]$$

$$-(f+1)f(f-2)/2 \cdot y[x_0 + 1] + (f+1)f(f-1)/6 \cdot y[x_0 + 2]$$

where $f = x - x_0$ is the fractional part of the index. For tables with 4 or more points, doubling the number of points on the table tends to improve the RMS error by 24 dB. Table 2.1 shows the calculated RMS error for sinusoids at various periods for 1, 2, and 4 point interpolation. (A slightly different quantity is measured in [Moo90, p.164]. There, the errors in amplitude and phase are also added in, yielding slightly more pessimistic results. See also [Har87].)

The allowable input domain for table lookup depends on the number of points of interpolation. In general, when using k-point interpolation into a table with N points, the input may range over an interval of $N + 1 - k$ points. If $k = 1$ (i.e., no interpolation at all), the domain is from 0 to N (including the endpoint at 0 but excluding the one at N) assuming input values are truncated (as is done for non-interpolated table lookup in Pd). The domain is from -1/2 to $N - 1/2$ if, instead, we round the input to the nearest integer instead of interpolating. In either case, the domain stretches over a length of N points.

For two-point interpolation, the input must lie between the first and last points, that is, between 0 and $N - 1$. So the N points suffice to define the function over a domain of length $N - 1$. For four-point interpolation, we cannot get values for inputs between 0 and 1 (not having the required two points to the left of the input) and neither can we for the space between the last two points ($N-2$ and $N-1$). So in this case the domain reaches from 1 to $N-2$ and has length $N-3$.

Periodic waveforms stored in wavetables require special treatment at the ends of the table. For example, suppose we wish to store a pure sinusoid of length N. For non-interpolating table lookup, it suffices to set, for example,

$$x[n] = \cos(2\pi n/N),\ n = 0, \ldots, N-1$$

For two-point interpolation, we need $N+1$ points:

$$x[n] = \cos(2\pi n/N),\ n = 0, \ldots, N$$

In other words, we must repeat the first ($n = 0$) point at the end, so that the last segment from $N-1$ to N reaches back to the beginning value.

For four-point interpolation, the cycle must be adjusted to start at the point $n = 1$, since we can't get properly interpolated values out for inputs less than one. If, then, one cycle of the wavetable is arranged from 1 to N, we must supply extra points for 0 (copied from N), and also $N+1$ and $N+2$, copied from 1 and 2, to make a table of length $N+3$. For the same sinusoid as above, the table should contain:

$$x[n] = \cos(2\pi(n-1)/N),\ n = 0, \ldots, N+2$$

2.6 Examples

Wavetable oscillator

Example B01.wavetables.pd, shown in Figure 2.12, implements a wavetable oscillator, which plays back from a wavetable named "table10". Two new Pd primitives are shown here. First is the wavetable itself, which appears at right in the figure. You can "mouse" on the wavetable to change its shape and hear the sound change as a result. Not shown in the figure but demonstrated in the patch is Pd's facility for automatically calculating wavetables with specified partial amplitudes, which is often preferable to drawing waveforms by hand. You can also read and write tables to (text or sound) files for interchanging data with other programs. The other novelty is an object class:

$\boxed{\text{tabosc4} \sim}$: a wavetable oscillator. The "4" indicates that this class uses 4-point (cubic) interpolation. In the example, the table's name, "table10", is specified as a creation argument to the tabosc4~ object. (You can also

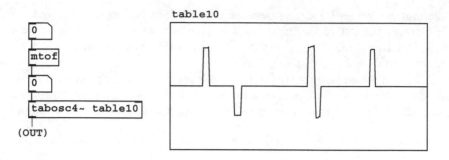

Figure 2.12: A wavetable oscillator: B01.wavetables.pd.

switch between wavetables dynamically by sending appropriate messages to the object.)

Wavetables used by **tabosc4~** must always have a period equal to a power of two; but as shown above, the wavetable must have three extra points wrapped around the ends. Allowable table lengths are thus of the form $2^m + 3$, such as 131, 259, 515, etc.

Wavetable oscillators are not limited to use as audio oscillators. Patch B02.wavetable.FM.pd (not pictured here) uses a pair of wavetable oscillators in series. The first one's output is used as the input of the second one, and thus controls its frequency which changes periodically in time.

Wavetable lookup in general

The **tabosc4~** class, while handy and efficient, is somewhat specialized and for many of the applications described in this chapter we need something more general. Example B03.tabread4.pd (Figure 2.13) demonstrates the timbre stretching technique discussed in Section 2.4. This is a simple example of a situation where **tabosc4~** would not have sufficed. There are new classes introduced here:

tabread4 ~ : wavetable lookup. As in **tabosc4~** the table is read using 4-point interpolation. But whereas **tabosc4~** takes a frequency as input and automatically reads the waveform in a repeating pattern, the simpler **tabread4~** expects the table lookup index as input. If you want to use it to do something repetitious, as in this example, the input itself has to be a repeating waveform. Like **tabosc4~** (and all the other table reading and writing objects), you can send messages to select which table to use.

tabwrite ~ : record an audio signal into a wavetable. In this example the

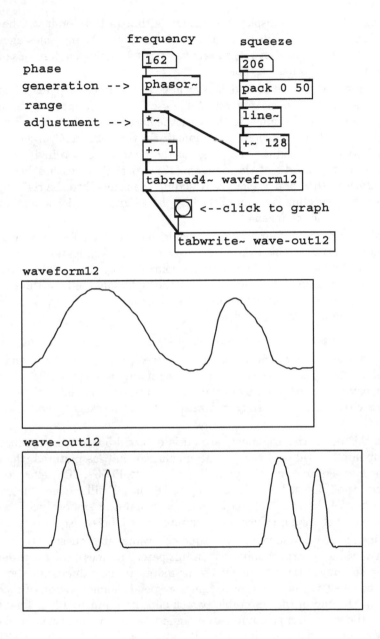

Figure 2.13: A wavetable oscillator with variable duty cycle: B03.tabread4.pd.

`tabwrite~` is used to display the output (although later on it will be used for all sorts of other things.) Whenever it receives a "bang" message from the pushbutton icon above it, `tabwrite~` begins writing successive samples of its input to the named table.

Example B03.tabread4.pd shows how to combine a `phasor~` and a `tabread4~` object to make a wavetable oscillator. The `phasor~`'s output ranges from 0 to 1 in value. In this case the input wavetable, named "waveform12", is 131 elements long. The domain for the `tabread4~` object is thus from 1 to 129. To adjust the range of the `phasor~` accordingly, we multiply it by the length of the domain (128) so that it reaches between 0 and 128, and then add 1, effectively sliding the interval to the right by one point. This rescaling is accomplished by the `*~` and `+~` objects between the `phasor~` and the `tabread4~`.

With only these four boxes we would have essentially reinvented the `tabosc4~` class. In this example, however, the multiplication is not by a constant 128 but by a variable amount controlled by the "squeeze" parameter. The function of the four boxes at the right hand side of the patch is to supply the `*~` object with values to scale the `phasor~` by. This makes use of one more new object class:

| pack |: compose a list of two or more elements. The creation arguments establish the number of arguments, their types (usually numbers) and their initial values. The inlets (there will be as many as you specified creation arguments) update the values of the message arguments, and, if the leftmost inlet is changed (or just triggered with a "bang" message), the message is output.

In this patch the arguments are initially 0 and 50, but the number box will update the value of the first argument, so that, as pictured, the most recent message to leave the `pack` object was "206 50". The effect of this on the `line~` object below is to ramp to 206 in 50 milliseconds; in general the output of the `line~` object is an audio signal that smoothly follows the sporadically changing values of the number box labeled "squeeze".

Finally, 128 is added to the "squeeze" value; if "squeeze" takes non-negative values (as the number box in this patch enforces), the range-setting multiplier ranges the phasor by 128 or more. If the value is greater than 128, the effect is that the rescaled phasor spends some fraction of its cycle stuck at the end of the wavetable (which clips its input to 129). The result is that the waveform is scanned over some fraction of the cycle. As shown, the waveform is squeezed into $128/(128+206)$ of the cycle, so the spectrum is stretched by a factor of about $1/2$.

For simplicity, this patch is subtly different from the example of Section 2.4 in that the waveforms are squeezed toward the beginning of each cycle and not toward the middle. This has the effect of slightly changing the phase of the various partials of the waveform as it is stretched and squeezed; if the squeezing factor changes quickly, the corresponding phase drift will sound like a slight wavering in pitch. This can be avoided by using a slightly more complicated arrangement: subtract 1/2 from the phasor˜, multiply it by 128 or more, and then add 65 instead of one.

Using a wavetable as a sampler

Example B04.sampler.pd (Figure 2.14) shows how to use a wavetable as a sampler. In this example the index into the sample (the wavetable) is controlled by mousing on a number box at top. A convenient scaling for

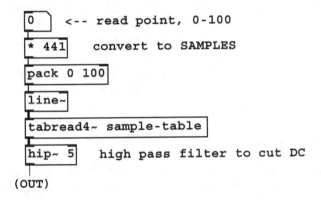

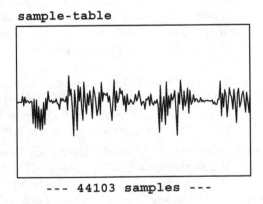

Figure 2.14: A sampler with mouse-controlled index: B04.sampler.pd.

the number box is hundredths of a second; to convert to samples (as the
input of `tabread4~` requires) we multiply by 44100 samples/sec times 0.01
sec to get 441 samples per unit, before applying `pack` and `line~` in much
the same way as they were used in the previous example. The transposition
you hear depends on how quickly you mouse up and down. This example
has introduced one new object class:

| hip ~ | : simple high-pass (low-cut) filter. The creation argument gives
the roll-off frequency in cycles per second. We use it here to eliminate
the constant (zero-frequency) output when the input sits in a single sample
(whenever you aren't actively changing the wavetable reading location with
the mouse). Filters are discussed in Chapter 8.

The `pack` and `line~` in this example are not included merely to make the
sound more continuous, but are essential to making the sound intelligible at
all. If the index into the wavetable lookup simply changed every time the
mouse moved a pixel (say, twenty to fifty times a second) the overwhelming
majority of samples would get the same index as the previous sample (the
other 44000+ samples, not counting the ones where the mouse moved.) So
the speed of precession would almost always be zero. Instead of changing
transpositions, you would hear 20 to 50 cycles-per-second grit. (Try it to
find out what that sounds like!)

Looping samplers

In most situations, you'll want a more automated way than moving the
mouse to specify wavetable read locations; for instance, you might want
to be able to play a sample at a steady transposition; you might have
several samples playing back at once (or other things requiring attention),
or you might want to switch quickly between samples or go to prearranged
locations. In the next few examples we'll develop an automated looping
sample reader, which, although only one of many possible approaches, is a
powerful and often-used one.

Patches B05.sampler.loop.pd and B06.sampler.loop.smooth.pd show how
to do this: the former in the simplest possible way and the latter (pictured
in Figure 2.15, part a) incorporating a second waveshape to envelope the
sound as described in Section 2.3. One new object class is introduced here:

| cos ~ | : compute the cosine of 2π times the input signal (so that 0 to 1
makes a whole cycle). Unlike the table reading classes in Pd, `cos~` handles
wraparound so that there is no range limitation on its input.

In Figure 2.15 (part a), a `phasor~` object supplies both indices into the
wavetable (at right) and phases for a half-cosine-shaped envelope function
at left. These two are multiplied, and the product is high-pass filtered

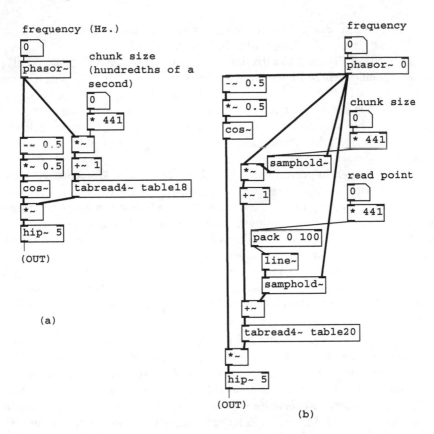

Figure 2.15: (a) a looping sampler with a synchronized envelope (B06.sampler.loop.smooth.pd); (b) the same, but with a control for read location (B08.sampler.nodoppler.pd).

and output. Reading the wavetable is straightforward; the phasor is multiplied by a "chunk size" parameter, added to 1, and used as an index to tabread4~. The chunk size parameter is multiplied by 441 to convert it from hundredths of a second to samples. This corresponds exactly to the block diagram shown in Figure 2.5, with a segment location of 1. (The segment location can't be 0 because 1 is the minimum index for which tabread4~ works.)

The left-hand signal path in the example corresponds to the enveloping wavetable lookup technique shown in Figure 2.7. Here the sawtooth wave is adjusted to the range (-1/4, 1/4) (by subtracting and multiplying by 0.5), and then sent to `cos~`. This reads the cosine function in the range $(-\pi/2, \pi/2)$, thus giving only the positive half of the waveform.

Part (b) of Figure 2.15 introduces a third parameter, the "read point", which specifies where in the sample the loop is to start. (In part (a) we always started at the beginning.) The necessary change is simple enough: add the "read point" control value, in samples, to the wavetable index and proceed as before. To avoid discontinuities in the index we smooth the read point value using `pack` and `line~` objects, just as we did in the first sampler example (Figure 2.14).

This raises an important, though subtle, issue. The Momentary Transposition Formula (Page 34) predicts that, as long as the chunk size and read point aren't changing in time, the transposition is just the frequency times the chunk size (as always, using appropriate units; Hertz and seconds, for example, so that the product is dimensionless). However, varying the chunk size and read point in time will affect the momentary transposition, often in very noticeable ways, as can be heard in Example B07.sampler.scratch.pd. Example B08.sampler.nodoppler.pd (the one shown in the figure) shows one possible way of controlling this effect, while introducing a new object class:

| samphold ~ |: a sample and hold unit. (This will be familiar to analog synthesizer users, but with a digital twist; for more details see Section 3.7.) This stores a single sample of the left-hand-side input and outputs it repeatedly, until caused by the right-hand-side input (also a digital audio signal, called the *trigger*) to overwrite the stored sample with a new one—again from the left-hand-side input. The unit acquires a new sample whenever the trigger's numerical value falls from one sample to the next. This is designed to be easy to pair with `phasor~` objects, to facilitate triggering on phase wraparounds.

Example B08.sampler.nodoppler.pd uses two `samphold~` objects to update the values of the chunk size and read point, exactly when the `phasor~` wraps around, at which moments the cosine envelope is at zero so the effect of the instantaneous changes can't be heard. In this situation we can apply the simpler Transposition Formula for Looping Wavetables to relate frequency, chunk size, and transposition. This is demonstrated in Example B09.sampler.transpose.pd (not shown).

Overlapping sample looper

As described in Section 2.3, it is sometimes desirable to use two or more overlapping looping samplers to produce a reasonably continuous sound without having to envelope too sharply at the ends of the loop. This is especially likely in situations where the chunk that is looped is short, a tenth of a second or less. Example B10.sampler.overlap.pd, shown in Figure 2.16 (part a), realizes two looping samplers a half-cycle out of phase from each other. New object classes are:

loadbang : output a "bang" message on load. This is used in this patch to make sure the division of transposition by chunk size will have a valid transposition factor in case "chunk size" is moused on first.

expr : evaluate mathematical expressions. Variables appear as $f1, $f2, and so on, corresponding to the object's inlets. Arithmetic operations are allowed, with parentheses for grouping, and many library functions are supplied, such as exponentiation, which shows up in this example as "pow" (the power function).

wrap ~ : wrap to the interval from 0 to 1. So, for instance, 1.2 becomes 0.2; 0.5 remains unchanged; and -0.6 goes to 0.4.

send ~ , s ~ , receive ~ , r ~ : signal versions of send and receive. An audio signal sent to a send~ object appears at the outlets of any and all receive~ objects of the same name. Unlike send and receive, you may not have more than one send~ object with the same name (in that connection, see the throw~ and catch~ objects).

In the example, part of the wavetable reading machinery is duplicated, using identical calculations of "chunk-size-samples" (a message stream) and "read-pt" (an audio signal smoothed as before). However, the "phase" audio signal, in the other copy, is replaced by "phase2". The top part of the figure shows the calculation of the two phase signals: the first one as the output of a phasor~ object, and the second by adding 0.5 and wrapping, thereby adding 0.5 cycles (π radians) to the phase. The two phase signals are each used, with the same range adjustments as before, to calculate indices into the wavetable and the cos~ object, and to control the two samphold~ objects. Finally, the results of the two copies are added for output.

Automatic read point precession

Example B11.sampler.rockafella.pd, shown in part (b) of Figure 2.16, adapts the ideas shown above to a situation where the read point is computed automatically. Here we precess the read-point through the sample in a loop, permitting us to speed up or slow down the playback independently of the transposition.

This example addresses a weakness of the preceding one, which is that, if the relative precession speed is anywhere near one (i.e., the natural speed of listening to the recorded wavetable), and if there is not much transposition

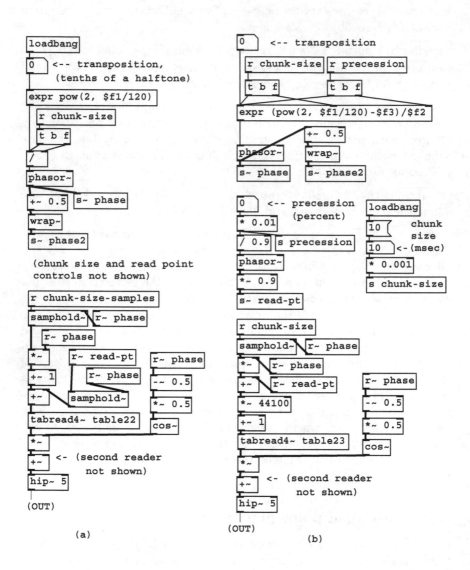

Figure 2.16: (a) two overlapped looping samplers (B10.sampler.overlap.pd); (b) the same, but with a phasor-controlled read point (B11.sampler.rockafella.pd).

either, it becomes preferable to use larger grains and lower the frequency of repetition accordingly (keeping the product constant to achieve the desired transposition.) However, if the grain size is allowed to get large, it is no longer convenient to quantize control changes at phase wrappings, because they might be too far apart to allow for a reasonable response time to control changes.

In this patch we remove the `samphold~` object that had controlled the read point (but we leave in the one for chunk size which is much harder to change in mid-loop). Instead, we use the (known) rate of precession of the read point to correct the sawtooth frequency, so that we maintain the desired transposition. It turns out that, when transposition factor and precession are close to each other (so that we are nearly doing the same thing as simple speed change) the frequency will drop to a value close to zero, so we will have increased the naturalness of the result at the same time.

In this patch we switch from managing read points, chunk sizes, etc., in samples and use seconds instead, converting to samples (and shifting by one) only just before the `tabread4~` object. The wavetable holds one second of sound, and we'll assume here that the nominal chunk size will not exceed 0.1 second, so that we can safely let the read point range from 0 to 0.9; the "real" chunk size will vary, and can become quite large, because of the moving read pointer.

The precession control sets the frequency of a phasor of amplitude 0.9, and therefore the precession must be multiplied by 0.9 to set the frequency of the phasor (so that, for a precession of one for instance, the amplitude and frequency of the read point are both 0.9, so that the slope, equal to amplitude over frequency, is one). The output of this is named "read-pt" as before, and is used by both copies of the wavetable reader.

The precession p and the chunk size c being known, and if we denote the frequency of the upper (original) `phasor~` by f, the transposition factor is given by:

$$t = p + cf$$

and solving for f gives:

$$f = \frac{t - p}{c} = \frac{2^{h/12} - p}{c}$$

where h is the desired transposition in half-steps. This is the formula used in the `expr` object.

Exercises

1. If a wavetable with 1000 samples is played back at unit transposition, at a sample rate of 44100 Hertz, how long does the resulting sound last?

2. A one-second wavetable is played back in 0.5 seconds. By what interval is the sound transposed?

3. Still assuming a one-second wavetable, if we play it back periodically (in a loop), at how many Hertz should we loop the wavetable to transpose the original sound upward one half-step?

4. We wish to play a wavetable (recorded at $R = 44100$), looping ten times per second, so that the original sound stored in the wavetable is transposed up a perfect fifth (see Page 14). How large a segment of the wavetable, in samples, should be played back?

5. Suppose you wish to use waveform stretching on a wavetable that holds a periodic waveform of period 100. You wish to hear the untransposed spectrum at a period of 200 samples. By what duty factor should you squeeze the waveform?

6. The first half of a wavetable contains a cycle of a sinusoid of peak amplitude one. The second half contains zeros. What is the strength of the second partial of the wavetable?

7. A sinusoid is stored in a wavetable with period 4 so that the first four elements are 0, 1, 0, and -1, corresponding to indices 0, 1, 2, and 3. What value do we get for an input of 1.5: (a) using 2-point interpolation? (b) using 4-point interpolation? (c) What's the value of the original sinusoid there?

8. If a wavetable's contents all fall between -1 and 1 in value, what is the range of possible outputs of wavetable lookup using 4-point interpolation?

Chapter 3

Audio and Control Computations

3.1 The Sampling Theorem

So far we have discussed digital audio signals as if they were capable of describing any function of time, in the sense that knowing the values the function takes on the integers should somehow determine the values it takes between them. This isn't really true. For instance, suppose some function f (defined for real numbers) happens to attain the value 1 at all integers:

$$f(n) = 1, \quad n = \ldots, -1, 0, 1, \ldots$$

We might guess that $f(t) = 1$ for all real t. But perhaps f happens to be one for integers and zero everywhere else—that's a perfectly good function too, and nothing about the function's values at the integers distinguishes it from the simpler $f(t) = 1$. But intuition tells us that the constant function is in the *spirit* of digital audio signals, whereas the one that hides a secret between the samples isn't. A function that is "possible to sample" should be one for which we can use some reasonable interpolation scheme to deduce its values on non-integers from its values on integers.

It is customary at this point in discussions of computer music to invoke the famous *Nyquist theorem*. This states (roughly speaking) that if a function is a finite or even infinite combination of sinusoids, none of whose angular frequencies exceeds π, then, theoretically at least, it is fully determined by the function's values on the integers. One possible way of reconstructing the function would be as a limit of higher- and higher-order polynomial interpolation.

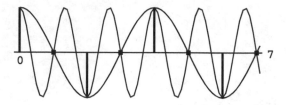

Figure 3.1: Two real sinusoids, with angular frequencies $\pi/2$ and $3\pi/2$, showing that they coincide at integers. A digital audio signal can't distinguish between the two.

The angular frequency π, called the *Nyquist frequency*, corresponds to $R/2$ cycles per second if R is the sample rate. The corresponding period is two samples. The Nyquist frequency is the best we can do in the sense that any real sinusoid of higher frequency is equal, at the integers, to one whose frequency is lower than the Nyquist, and it is this lower frequency that will get reconstructed by the ideal interpolation process. For instance, a sinusoid with angular frequency between π and 2π, say $\pi + \omega$, can be written as

$$\cos((\pi + \omega)n + \phi) = \cos((\pi + \omega)n + \phi - 2\pi n)$$

$$= \cos((\omega - \pi)n + \phi)$$

$$= \cos((\pi - \omega)n - \phi)$$

for all integers n. (If n weren't an integer the first step would fail.) So a sinusoid with frequency between π and 2π is equal, on the integers at least, to one with frequency between 0 and π; you simply can't tell the two apart. And since any conversion hardware should do the "right" thing and reconstruct the lower-frequency sinusoid, any higher-frequency one you try to synthesize will come out your speakers at the wrong frequency—specifically, you will hear the unique frequency between 0 and π that the higher frequency lands on when reduced in the above way. This phenomenon is called *foldover*, because the half-line of frequencies from 0 to ∞ is folded back and forth, in lengths of π, onto the interval from 0 to π. The word *aliasing* means the same thing. Figure 3.1 shows that sinusoids of angular frequencies $\pi/2$ and $3\pi/2$, for instance, can't be distinguished as digital audio signals.

We conclude that when, for instance, we're computing values of a Fourier series (Page 13), either as a wavetable or as a real-time signal, we had better leave out any sinusoid in the sum whose frequency exceeds π. But the picture in general is not this simple, since most techniques other than additive synthesis don't lead to neat, band-limited signals (ones whose components stop at some limited frequency). For example, a sawtooth wave of frequency ω, of the form put out by Pd's `phasor~` object but considered as a continuous function $f(t)$, expands to:

$$f(t) = \frac{1}{2} \ \ \frac{1}{\pi}\left(\sin(\omega t) + \frac{\sin(2\omega t)}{2} + \frac{\sin(3\omega t)}{3} + \cdots\right)$$

which enjoys arbitrarily high frequencies; and moreover the hundredth partial is only 40 dB weaker than the first one. At any but very low values of ω, the partials above π will be audibly present—and, because of foldover, they will be heard at incorrect frequencies. (This does not mean that one shouldn't use sawtooth waves as phase generators—the wavetable lookup step magically corrects the sawtooth's foldover—but one should think twice before using a sawtooth wave itself as a digital sound source.)

Many synthesis techniques, even if not strictly band-limited, give partials which may be made to drop off more rapidly than $1/n$ as in the sawtooth example, and are thus more forgiving to work with digitally. In any case, it is always a good idea to keep the possibility of foldover in mind, and to train your ears to recognize it.

The first line of defense against foldover is simply to use high sample rates; it is a good practice to systematically use the highest sample rate that your computer can easily handle. The highest practical rate will vary according to whether you are working in real time or not, CPU time and memory constraints, and/or input and output hardware, and sometimes even software-imposed limitations.

A very non-technical treatment of sampling theory is given in [Bal03]. More detail can be found in [Mat69, pp. 1-30].

3.2 Control

So far we have dealt with audio signals, which are just sequences $x[n]$ defined for integers n, which correspond to regularly spaced points in time. This is often an adequate framework for describing synthesis techniques, but real electronic music applications usually also entail other computations which have to be made at irregular points in time. In this section we'll develop a framework for describing what we will call *control* computations. We will always require that any computation correspond to a specific *logical time*. The logical time controls which sample of audio output will be the first to reflect the result of the computation.

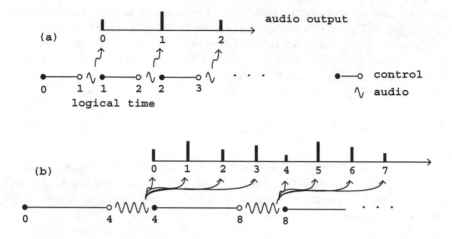

Figure 3.2: Timeline for digital audio and control computation: (a) with a block size of one sample; (b) with a block size of four samples.

In a non-real-time system (such as Csound in its classical form), this means that logical time proceeds from zero to the length of the output soundfile. Each "score card" has an associated logical time (the time in the score), and is acted upon once the audio computation has reached that time. So audio and control calculations (grinding out the samples and handling note cards) are each handled in turn, all in increasing order of logical time.

In a real-time system, logical time, which still corresponds to the time of the next affected sample of audio output, is always slightly in advance of *real time*, which is measured by the sample that is actually leaving the computer. Control and audio computations still are carried out in alternation, sorted by logical time.

The reason for using logical time and not real time in computer music computations is to keep the calculations independent of the actual execution time of the computer, which can vary for a variety of reasons, even for two seemingly identical calculations. When we are calculating a new value of an audio signal or processing some control input, real time may pass but we require that the logical time stay the same through the whole calculation, as if it took place instantaneously. As a result of this, electronic music computations, if done correctly, are deterministic: two runs of the same real-time or non-real-time audio computation, each having the same inputs, should have identical results.

Figure 3.2 (part a) shows schematically how logical time and sample computation are lined up. Audio samples are computed at regular periods (marked as wavy lines), but before the calculation of each sample we do all the control calculations that might affect it (marked as straight line segments). First we do the control computations associated with logical times starting at zero, up to but not including one; then we compute the first audio sample (of index zero), at logical time one. We then do all control calculations up to but not including logical time 2, then the sample of index one, and so on. (Here we are adopting certain conventions about labeling that could be chosen differently. For instance, there is no fundamental reason control should be pictured as coming "before" audio computation but it is easier to think that way.)

Part (b) of the figure shows the situation if we wish to compute the audio output in blocks of more than one sample at a time. Using the variable B to denote the number of elements in a block (so $B = 4$ in the figure), the first audio computation will output samples $0, 1, ...B - 1$ all at once in a block computed at logical time B. We have to do the relevant control computations for all B periods of time in advance. There is a delay of B samples between logical time and the appearance of audio output.

Most computer music software computes audio in blocks. This is done to increase the efficiency of individual audio operations (such as Csound's unit generators and Max/MSP and Pd's tilde objects). Each unit generator or tilde object incurs overhead each time it is called, equal to perhaps twenty times the cost of computing one sample on average. If the block size is one, this means an overhead of 2,000%; if it is sixty-four (as in Pd by default), the overhead is only some 30%.

3.3 Control Streams

Control computations may come from a variety of sources, both internal and external to the overall computation. Examples of internally engendered control computations include sequencing (in which control computations must take place at pre-determined times) or feature detection of the audio output (for instance, watching for zero crossings in a signal). Externally engendered ones may come from input devices such as MIDI controllers, the mouse and keyboard, network packets, and so on. In any case, control computations may occur at irregular intervals, unlike audio samples which correspond to a steadily ticking sample clock.

We will need a way of describing how information flows between control and audio computations, which we will base on the notion of a *control stream*. This is simply a collection of numbers—possibly empty—that appear as a result of control computations, whether regularly or irregularly spaced in logical time. The simplest possible control stream has no infor-

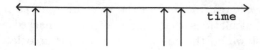

Figure 3.3: Graphical representation of a control stream as a sequence of points in time.

mation other than a *time sequence*:

$$\ldots, t[0], t[1], t[2], \ldots$$

Although the time values are best given in units of samples, their values aren't quantized; they may be arbitrary real numbers. We do require them to be sorted in nondecreasing order:

$$\cdots \leq t[0] \leq t[1] \leq t[2] \leq \cdots$$

Each item in the sequence is called an *event*.

Control streams may be shown graphically as in Figure 3.3. A number line shows time and a sequence of arrows points to the times associated with each event. The control stream shown has no data (it is a time sequence). If we want to show data in the control stream we will write it at the base of each arrow.

A *numeric control stream* is one that contains one number per time point, so that it appears as a sequence of ordered pairs:

$$\ldots, (t[0], x[0]), (t[1], x[1]), \ldots$$

where the $t[n]$ are the time points and the $x[n]$ are the signal's values at those times.

A numeric control stream is roughly analogous to a "MIDI controller", whose values change irregularly, for example when a physical control is moved by a performer. Other control stream sources may have higher possible rates of change and/or more precision. On the other hand, a time sequence might be a sequence of pedal hits, which (MIDI implementation notwithstanding) shouldn't be considered as having *values*, just *times*.

Numeric control streams are like audio signals in that both are just time-varying numeric values. But whereas the audio signal comes at a steady rate (and so the time values need not be specified per sample), the control stream comes unpredictably—perhaps evenly, perhaps unevenly, perhaps never.

Let us now look at what happens when we try to convert a numeric control stream to an audio signal. As before we'll choose a block size $B = 4$. We will consider as a control stream a square wave of period 5.5:

$$(2, 1), (4.75, 0), (7.5, 1), (10.25, 0), (13, 1), \ldots$$

and demonstrate three ways it could be converted to an audio signal. Figure 3.4 (part a) shows the simplest, fast-as-possible, conversion. Each audio sample of output simply reflects the most recent value of the control signal. So samples 0 through 3 (which are computed at logical time 4 because of the block size) are 1 in value because of the point (2, 1). The next four samples are also one, because of the two points, (4.75, 0) and (7.5, 1), the most recent still has the value 1.

Fast-as-possible conversion is most appropriate for control streams which do not change frequently compared to the block size. Its main advantages are simplicity of computation and the fastest possible response to changes. As the figure shows, when the control stream's updates are too fast (on the order of the block size), the audio signal may not be a good likeness of the sporadic one. (If, as in this case, the control stream comes at regular intervals of time, we can use the sampling theorem to analyze the result. Here the Nyquist frequency associated with the block rate R/B is lower than the input square wave's frequency, and so the output is aliased to a new frequency lower than the Nyquist frequency.)

Part (b) shows the result of nearest-sample conversion. Each new value of the control stream at a time t affects output samples starting from index $\lfloor t \rfloor$ (the greatest integer not exceeding t). This is equivalent to using fast-as-possible conversion at a block size of 1; in other words, nearest-sample conversion hides the effect of the larger block size. This is better than fast-as-possible conversion in cases where the control stream might change quickly.

Part (c) shows sporadic-to-audio conversion, again at the nearest sample, but now also using two-point interpolation to further increase the time accuracy. Conceptually we can describe this as follows. Suppose the value of the control stream was last equal to x, and that the next point is $(n+f, y)$, where n is an integer and f is the fractional part of the time value (so $0 \le f < 1$). The first point affected in the audio output will be the sample at index n. But instead of setting the output to y as before, we set it to

$$fx + (1 - f)y$$

in other words, to a weighted average of the previous and the new value, whose weights favor the new value more if the time of the sporadic value is earlier, closer to n. In the example shown, the transition from 0 to 1 at

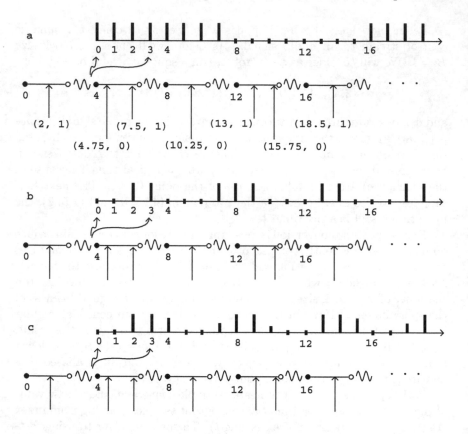

Figure 3.4: Three ways to change a control stream into an audio signal: (a) as fast as possible; (b) delayed to the nearest sample; (c) with two-point interpolation for higher delay accuracy.

time 2 gives

$$0 \cdot x + 1 \cdot y = 1$$

while the transition from 1 to 0 at time 4.75 gives:

$$0.75 \cdot x + 0.25 \cdot y = 0.75$$

This technique gives a still closer representation of the control signal (at least, the portion of it that lies below the Nyquist frequency), at the expense of more computation and slightly greater delay.

Numeric control streams may also be converted to audio signals using ramp functions to smooth discontinuities. This is often used when a control stream is used to control an amplitude, as described in Section 1.5. In

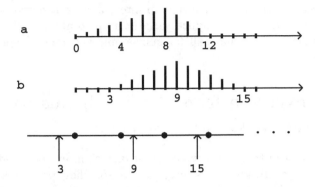

Figure 3.5: Line segment smoothing of numeric control streams: (a) aligned to block boundaries; (b) aligned to nearest sample.

general there are three values to specify to set a ramp function in motion: a start time and target value (specified by the control stream) and a target time, often expressed as a delay after the start time.

In such situations it is almost always accurate enough to adjust the start and ending times to match the first audio sample computed at a later logical time, a choice which corresponds to the fast-as-possible scenario above. Figure 3.5 (part a) shows the effect of ramping from 0, starting at time 3, to a value of 1 at time 9, immediately starting back toward 0 at time 15, with block size $B = 4$. The times 3, 9, and 15 are truncated to 0, 8, and 12, respectively.

In real situations the block size might be on the order of a millisecond, and adjusting ramp endpoints to block boundaries works fine for controlling amplitudes; reaching a target a fraction of a millisecond early or late rarely makes an audible difference. However, other uses of ramps are more sensitive to time quantization of endpoints. For example, if we wish to do something repetitively every few milliseconds, the variation in segment lengths will make for an audible aperiodicity.

For situations such as these, we can improve the ramp generation algorithm to start and stop at arbitrary samples, as shown in Figure 3.5 (part b), for example. Here the endpoints of the line segments line up exactly with the requested samples 3, 9, and 15. We can go even further and adjust for fractional samples, making the line segments touch the values 0 and 1 at exactly specifiable points on a number line.

For example, suppose we want to repeat a recorded sound out of a wavetable 100 times per second, every 441 samples at the usual sample rate. Rounding errors due to blocking at 64-sample boundaries could detune the playback by as much as a whole tone in pitch; and even rounding to one-sample boundaries could introduce variations up to about 0.2%, or three cents. This situation would call for sub-sample accuracy in sporadic-to-audio conversion.

3.4   Conversion from Audio Signals to Numeric Control Streams

We sometimes need to convert in the other direction, from an audio signal to a sporadic one. To go in this direction, we somehow provide a series of logical times (a time sequence), as well as an audio signal. For output we want a control stream combining the time sequence with values taken from the audio signal. We do this when we want to incorporate the signal's value as part of a control computation.

For example, we might be controlling the amplitude of a signal using a `line~` object as in Example A03.line.pd (Page 21). Suppose we wish to turn off the sound at a fixed rate of speed instead of in a fixed amount of time. For instance, we might want to re-use the network for another sound and wish to mute it as quickly as possible without audible artifacts; we probably can ramp it off in less time if the current amplitude is low than if it is high. To do this we must confect a message to the `line~` object to send it to zero in an amount of time we'll calculate on the basis of its current output value. This will require, first of all, that we "sample" the `line~` object's output (an audio signal) into a control stream.

The same issues of time delay and accuracy appear as for sporadic to audio conversion. Again there will be a tradeoff between immediacy and accuracy. Suppose as before that we are calculating audio in blocks of 4 samples, and suppose that at logical time 6 we want to look at the value of an audio signal, and use it to change the value of another one. As shown in Figure 3.2 (part b), the most recently calculated value of the signal will be for index 3 and the earliest index at which our calculation can affect a signal is 4. We can therefore carry out the whole affair with a delay of only one sample. However, we can't choose exactly *which* sample—the update can occur only at a block boundary.

As before, we can trade immediacy for increased time accuracy. If it matters exactly at which sample we carry out the audio-to-control-to-audio computation, we read the sample of index 2 and update the one at index 6. Then if we want to do the same thing again at logical time 7, we read from index 3 and update at index 7, and so on. In general, if the block size is B, and for any index n, we can always read the sample at index $n - B$ and affect the one at index n. There is thus a round-trip delay of B samples

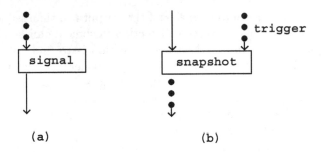

Figure 3.6: Conversion between control and audio: (a) control to signal; (b) signal to control by snapshots.

in going from audio to control to audio computation, which is the price incurred for being able to name the index n exactly.

If we wish to go further, to being able to specify a fraction of a sample, then (as before) we can use interpolation—at a slight further increase in delay. In general, as in the case of sporadic-to-audio conversion, in most cases the simplest solution is the best, but occasionally we have to do extra work.

3.5   Control Streams in Block Diagrams

Figure 3.6 shows how control streams are expressed in block diagrams, using control-to-signal and signal-to-control conversion as examples. Control streams are represented using dots (as opposed to audio signals which appear as solid arrows).

The *signal* block converts from a numeric control stream to an audio signal. The exact type of conversion isn't specified at this level of detail; in the Pd examples the choice of conversion operator will determine this.

The *snapshot* block converts from audio signals back to numeric control streams. In addition to the audio signal, a separate, control input is needed to specify the time sequence at which the audio signal is sampled.

3.6   Event Detection

Besides taking snapshots, a second mode of passing information from audio signals to control computations is *event detection*. Here we derive time information from the audio signal. An example is *threshold detection*, in which the input is an audio signal and the output is a time sequence. We'll consider the example of threshold detection in some detail here.

A typical reason to use threshold detection is to find out when some kind of activity starts and stops, such as a performer playing an instrument. We'll suppose we already have a continuous measure of activity in the form of an audio signal. (This can be done, for example, using an *envelope follower*). What we want is a pair of time sequences, one which marks times in which activity starts, and the other marking stops.

Figure 3.7 (part a) shows a simple realization of this idea. We assume the signal input is as shown in the continuous graph. A horizontal line shows the constant value of the threshold. The time sequence marked "onsets" contains one event for each time the signal crosses the threshold from below to above; the one marked "turnoffs" marks crossings in the other direction.

In many situations we will get undesirable onsets and turnoffs caused by small ripples in the signal close to the threshold. This is avoided by *debouncing*, which can be done in at least two simple ways. First, as shown in part (b) of the figure, we can set two thresholds: a high one for marking onsets, and a lower one for turnoffs. In this scheme the rule is that we only report the first onset after each turnoff, and, *vice versa*, we only report one turnoff after each onset. Thus the third time the signal crosses the high threshold in the figure, there is no reported onset because there was no turnoff since the previous one. (At startup, we act as if the most recent output was a turnoff, so that the first onset is reported.)

A second approach to filtering out multiple onsets and turnoffs, shown in part (c) of the figure, is to associate a *dead period* to each onset. This is a constant interval of time after each reported onset, during which we refuse to report more onsets or turnoffs. After the period ends, if the signal has dropped below the threshold in the meantime, we belatedly report a turnoff. Dead periods may also be associated with turnoffs, and the two time periods may have different values.

The two filtering strategies may be used separately or simultaneously. It is usually necessary to tailor the threshold values and/or dead times by hand to each specific situation in which thresholding is used.

Thresholding is often used as a first step in the design of higher-level strategies for arranging computer responses to audible cues from performers. A simple example could be to set off a sequence of pre-planned processes, each one to be set off by an onset of sound after a specified period of relative silence, such as you would see if a musician played a sequence of phrases separated by rests.

More sophisticated detectors (built on top of threshold detection) could detect continuous sound or silence within an expected range of durations, or sequences of quick alternation between playing and not playing, or

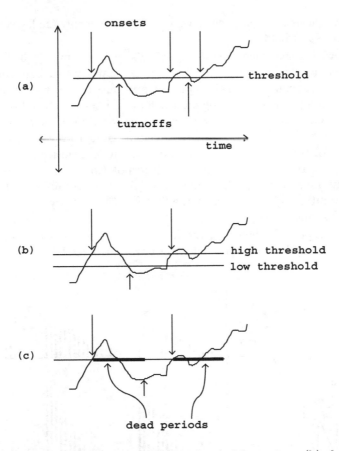

Figure 3.7: Threshold detection: (a) with no debouncing; (b) debounced using two threshold levels; (c) debounced using dead periods.

periods of time in which the percentage of playing time to rests is above or below a threshold, or many other possible features. These could set off predetermined reactions or figure in an improvisation.

3.7   Audio Signals as Control

From the tradition of analog synthesis comes an elegant, old-fashioned approach to control problems that can be used as an alternative to the control streams we have been concerned with so far in this chapter. Instead, or in addition to using control streams, we can use audio signals themselves to control the production of other audio signals. Two specific techniques from

analog synthesis lend themselves well to this treatment: analog sequencing
and sample-and-hold.

The analog sequencer [Str95, pp. 70-79] [Cha80, pp. 93,304-308] was
often used to set off a regularly or semi-regularly repeating sequence of
sounds. The sequencer itself typically put out a repeating sequence of
voltages, along with a trigger signal which pulsed at each transition between
voltages. One used the voltages for pitches or timbral parameters, and
the trigger to control one or more envelope generators. Getting looped
sequences of predetermined values in digital audio practice is as simple as
sending a **phasor~** object into a non-interpolating table lookup. If you
want, say, four values in the sequence, scale the **phasor~** output to take
values from 0 to 3.999... so that the first fourth of the cycle reads point 0
of the table and so on.

To get repeated triggering, the first step is to synthesize another sawtooth that runs in synchrony with the **phasor~** output but four times as

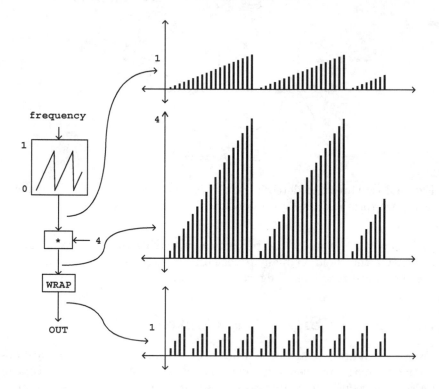

Figure 3.8: Multiplying and wrapping a sawtooth wave to generate a higher
frequency.

fast. This is done using a variant of the technique of Figure 2.8, in which we used an adder and a wraparound operator to get a desired phase shift. Figure 3.8 shows the effect of multiplying a sawtooth wave by an integer, then wrapping around to get a sawtooth at a multiple of the original frequency.

From there is is easy to get to a repeated envelope shape by wavetable lookup for example (using an interpolating table lookup this time, unlike the sequence voltages). All the waveform generation and altering techniques used for making pitched sounds can also be brought to use here.

The other standard control technique from analog synthesizer control is the sample and hold unit [Str95, pp. 80-83] [Cha80, p. 92]. This takes an incoming signal, picks out certain instantaneous values from it, and "freezes" those values for its output. The particular values to pick out are selected by a secondary, "trigger" input. At points in time specified by the trigger input a new, single value is taken from the primary input and is output continuously until the next time point, when it is replaced by a new value of the primary input.

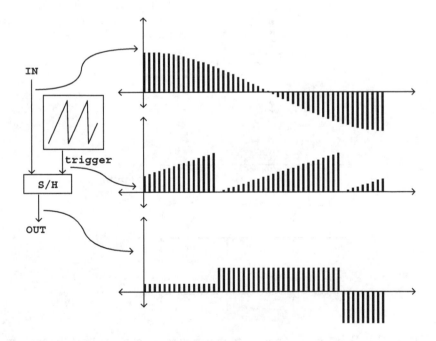

Figure 3.9: Sample and hold ("S/H"), using falling edges of the trigger signal.

In digital audio it is often useful to sample a new value on falling edges of the trigger signal, i.e., whenever the current value of the trigger signal is smaller than its previous value, as shown in Figure 3.9. This is especially convenient for use with a sawtooth trigger, when we wish to sample signals in synchrony with an oscillator-driven process. Pd's sample and hold object was previously introduced in the context of sampling (Example B08.sampler.nodoppler.pd, Page 53).

3.8   Operations on Control Streams

So far we've discussed how to convert between control streams and audio streams. In addition to this possibility, there are four types of operations you can perform on control streams to get other control streams. These control stream operations have no corresponding operations on audio signals. Their existence explains in large part why it is useful to introduce a whole control structure in parallel with that of audio signals.

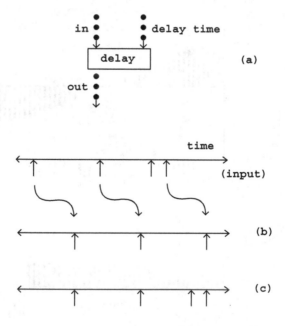

Figure 3.10: Delay as an operation on a control stream: (a) block diagram; (b) effect of a simple delay on a control stream; (c) effect of a compound delay.

The first type consists of *delay* operations, which offset the time values associated with a control stream. In real-time systems the delays can't be negative in value. A control stream may be delayed by a constant amount, or alternatively, you can delay each event separately by different amounts.

Two different types of delay are used in practice: *simple* and *compound*. Examples of each are shown in Figure 3.10. A simple delay acting on a control stream schedules each event, as it comes in, for a time in the future. However, if another event arrives at the input before the first event

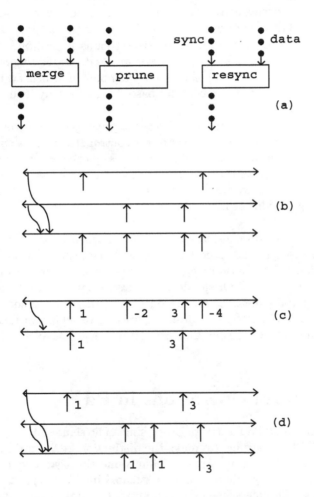

Figure 3.11: Operations on control streams (besides delay): (a) block diagrams; (b) merging; (c) pruning; (d) resynchronizing.

is output, the first event is forgotten in favor of the second. In a compound delay, each event at the input produces an output, even if other inputs arrive before the output appears.

A second operation on control steams is *merging*: taking two control streams and combining all the events into a new one. Figure 3.11 (part a) shows how this and the remaining operations are represented in block diagrams.

Part (b) of the figure shows the effect of merging two streams. Streams may contain more than one event at the same time. If two streams to be merged contain events at the same time, the merged stream contains them both, in a well-defined order.

A third type of operation on control streams is *pruning*. Pruning a control stream means looking at the associated data and letting only certain elements through. Part (c) shows an example, in which events (which each have an associated number) are passed through only if the number is positive.

Finally, there is the concept of *resynchronizing* one control stream to another, as shown in part (d). Here one control stream (the source) contributes values which are put onto the time sequence of a second one (the sync). The value given the output is always the most recent one from the source stream. Note that any event from the source may appear more than once (as suggested in the figure), or, on the other hand, it might not appear at all.

Again, we have to consider what happens when the two streams each contain an event at the same time. Should the sync even be considered as happening before the source (so that the output gets the value of the previous source event)? Or should the source event be considered as being first so that its value goes to the output at the same time? How this should be disambiguated is a design question, to which various software environments take various approaches. (In Pd it is controlled explicitly by the user.)

3.9   Control Operations in Pd

So far we have used Pd mostly for processing audio signals, although as early as Figure 1.10 we have had to make the distinction between Pd's notion of audio signals and of control streams: thin connections carry control streams and thick ones carry audio. Control streams in Pd appear as sequences of *messages*. The messages may contain data (most often, one or more numbers), or not. A *numeric control stream* (Section 3.3) appears as a (thin) connection that carries numbers as messages.

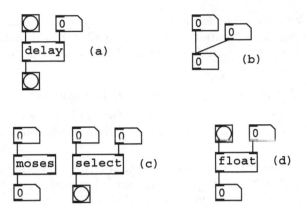

Figure 3.12: The four control operations in Pd: (a) delay; (b) merging; (c) pruning; (d) resynchronizing.

Messages not containing data make up *time sequences*. So that you can see messages with no data, in Pd they are given the (arbitrary) symbol "bang".

The four types of control operations described in the previous section can be expressed in Pd as shown in Figure 3.12. *Delays* are accomplished using two explicit delay objects:

del , delay : simple delay. You can specify the delay time in a creation argument or via the right inlet. A "bang" in the left inlet sets the delay, which then outputs "bang" after the specified delay in milliseconds. The delay is *simple* in the sense that sending a bang to an already set delay resets it to a new output time, canceling the previously scheduled one.

pipe : compound delay. Messages coming in the left inlet appear on the output after the specified delay, which is set by the first creation argument. If there are more creation arguments, they specify one or more inlets for numeric or symbolic data the messages will contain. Any number of messages may be stored by pipe simultaneously, and messages may be reordered as they are output depending on the various delay times given for them.

Merging of control streams in Pd is accomplished not by explicit objects but by Pd's connection mechanism itself. This is shown in part (b) of the figure with number boxes as an example. In general, whenever more than one connection is made to a control inlet, the control streams are merged.

Pd offers several objects for *pruning* control streams, of which two are shown in part (c) of the figure:

| moses | : prune for numeric range. Numeric messages coming in the left inlet appear on the left output if they are smaller than a threshold value (set by a creation argument or by the right inlet), and out the right inlet otherwise.

| select |, | sel | : prune for specific numbers. Numeric messages coming in the left inlet produce a "bang" on the output only if they match a test value exactly. The test value is set either by creation argument or from the right inlet.

Finally, Pd takes care of *resynchronizing* control streams implicitly in its connection mechanism, as illustrated by part (d) of the figure. Most objects with more than one inlet synchronize all other inlets to the leftmost one. So the `float` object shown in the figure resynchronizes its right-hand-side inlet (which takes numbers) to its left-hand-side one. Sending a "bang" to the left inlet outputs the most recent number `float` has received beforehand.

3.10   Examples

Sampling and foldover

Example C01.nyquist.pd (Figure 3.13, part a) shows an oscillator playing a wavetable, sweeping through frequencies from 500 to 1423. The wavetable consists of only the 46th partial, which therefore varies from 23000 to 65458 Hertz. At a sample rate of 44100 these two frequencies theoretically sound at 21100 and 21358 Hertz, but sweeping from one to the other folds down through zero and back up.

Two other waveforms are provided to show the interesting effects of beating between partials which, although they "should" have been far apart, find themselves neighbors through foldover. For instance, at 1423 Hertz, the second harmonic is 2846 Hertz whereas the 33rd harmonic sounds at 1423*33-44100 = 2859 Hertz—a rude dissonance.

Other less extreme examples can still produce audible foldover in less striking forms. Usually it is still objectionable and it is worth learning to hear it. Example C02.sawtooth-foldover.pd (not pictured here) demonstrates this for a sawtooth (the `phasor~` object). For wavetables holding audio recordings, interpolation error can create extra foldover. The effects of this can vary widely; the sound is sometimes described as "crunchy" or "splattering", depending on the recording, the transposition, and the interpolation algorithm.

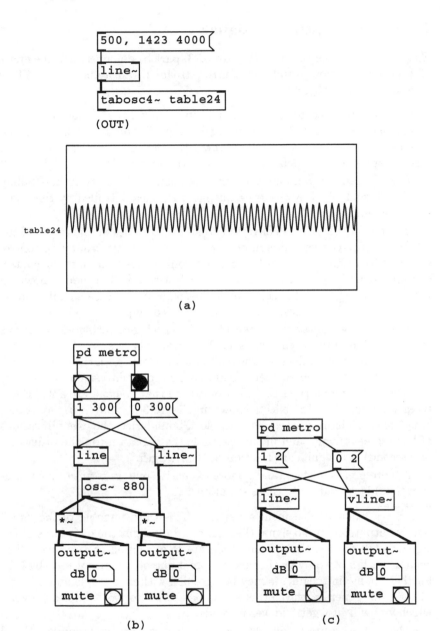

Figure 3.13: (a) sending an oscillator over the Nyquist frequency; (b) zipper noise from the `line` (control) object; (c) the `line~` and `vline~` objects compared.

Converting controls to signals

Example C03.zipper.noise.pd (Figure 3.13, part b) demonstrates the effect of converting a slowly-updated control stream to an audio signal. This introduces a new object:

| line |: a ramp generator with control output. Like `line~`, `line` takes pairs of numbers as (target, time) pairs and ramps to the target in the given amount of time; however, unlike `line~`, the output is a numeric control stream, appearing, by default, at 20 msec time intervals.

In the example you can compare the sound of the rising and falling amplitude controlled by the `line` output with one controlled by the audio signal generated by `line~`.

The output of `line` is converted to an audio signal at the input of the `*~` object. The conversion is implied here by connecting a numeric control stream into a signal inlet. In Pd, implicit conversions from numeric control streams to audio streams is done in the fast-as-possible mode shown in Figure 3.4 (part a). The `line` output becomes a staircase signal with 50 steps per second. The result is commonly called "zipper noise".

Whereas the limitations of the `line` object for generating audio signals were clearly audible even over such long time periods as 300 msec, the signal variant, `line~`, does not yield audible problems until the time periods involved become much shorter. Example C04.control.to.signal.pd (Figure 3.13, part c) demonstrates the effect of using `line~` to generate a 250 Hertz triangle wave. Here the effects shown in Figure 3.5 come into play. Since `line~` always aligns line segments to block boundaries, the exact durations of line segments vary, and in this example the variation (on the order of a millisecond) is a significant fraction of their length.

A more precise object (and a more expensive one, in terms of computation time) is provided for these situations:

| vline ~ |: exact line segment generator. This third member of the "line" family outputs an audio signal (like `line~`), but aligns the endpoints of the signal to the desired time points, accurate to a fraction of a sample. (The accuracy is limited only by the floating-point numerical format used by Pd.) Further, many line segments may be specified withing a single audio block; `vline~` can generate waveforms at periods down to two samples (beyond which you will just get foldover instead).

The `vline~` object can also be used for converting numeric control streams to audio streams in the nearest-sample and two-point-interpolation modes as shown in Figure 3.4 (parts b and c). To get nearest-sample conversion, simply give `vline~` a ramp time of zero. For linear interpolation, give it a ramp time of one sample (0.0227 msec if the sample rate is 44100 Hertz).

Non-looping wavetable player

One application area requiring careful attention to the control stream/audio signal boundary is sampling. Until now our samplers have skirted the issue by looping perpetually. This allows for a rich variety of sound that can be accessed by making continuous changes in parameters such as loop size and envelope shape. However, many uses of sampling require the internal features of a wavetable to emerge at predictable, synchronizable moments in time. For example, recorded percussion sounds are usually played from the beginning, are not often looped, and are usually played in a determined time relationship with the rest of the music.

In this situation, control streams are better adapted than audio signals as triggers. Example C05.sampler.oneshot.pd (Figure 3.14) shows one possible way to accomplish this. The four tilde objects at bottom left form the signal processing network for playback. One **vline~** object generates a phase signal (actually just a table lookup index) to the **tabread4~** object; this replaces the **phasor~** of Example B03.tabread4.pd (Page 49) and its derivatives.

The amplitude of the output of **tabread4~** is controlled by a second **vline~** object, in order to prevent discontinuities in the output in case a new event is started while the previous event is still playing. The "cutoff" **vline~** object ramps the output down to zero (whether or not it is playing) so that, once the output is zero, the index of the wavetable may be changed discontinuously.

In order to start a new "note", first, the "cutoff" **vline~** object is ramped to zero; then, after a delay of 5 msec (at which point **vline~** has reached zero) the phase is reset. This is done with two messages: first, the phase is set to 1 (with no time value so that it jumps to 1 with no ramping). The value "1" specifies the first readable point of the wavetable, since we are using 4-point interpolation. Second, in the same message box, the phase is ramped to 441,000,000 over a time period of 10,000,000 msec. (In Pd, large numbers are shown using exponential notation; these two appear as 4.41e+08 and 1e+07.) The quotient is 44.1 (in units per millisecond) giving a transposition of one. The upper **vline~** object (which generates the phase) receives these messages via the "r phase" object above it.

The example assumes that the wavetable is ramped smoothly to zero at either end, and the bottom right portion of the patch shows how to record such a wavetable (in this case four seconds long). Here a regular (and computationally cheaper) **line~** object suffices. Although the wavetable should be at least 4 seconds long for this to work, you may record shorter wavetables simply by cutting the **line~** object off earlier. The only caveat is that, if you are simultaneously reading and writing from the same wavetable, you

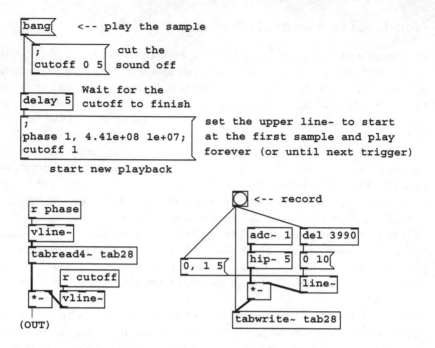

Figure 3.14: Non-looping sampler.

should avoid situations where read and write operations attack the same portion of the wavetable at once.

The vline~ objects surrounding the tabread4~ were chosen over line~ because the latter's rounding of breakpoints to the nearest block boundary (typically 1.45 msec) can make for audible aperiodicities in the sound if the wavetable is repeated more than 10 or 20 times per second, and would prevent you from getting a nice, periodic sound at higher rates of repetition.

We will return to vline~-based sampling in the next chapter, to add transposition, envelopes, and polyphony.

Signals to controls

Example C06.signal.to.control.pd (not pictured) demonstrates conversion from audio signals back to numeric control streams, via a new tilde object introduced here.

snapshot ~ : convert audio signal to control messages. This always gives the most recently computed audio sample (fast-as-possible conversion), so the exact sampling time varies by up to one audio block.

It is frequently desirable to sense the audio signal's amplitude rather than peek at a single sample; Example C07.envelope.follower.pd (also not pictured) introduces another object which does this:

env ~ : RMS envelope follower. Outputs control messages giving the short-term RMS amplitude (in decibels) of the incoming audio signal. A creation argument allows you to select the number of samples used in the RMS computation; smaller numbers give faster (and possibly less stable) output.

Analog-style sequencer

Example C08.analog.sequencer.pd (Figure 3.15) realizes the analog sequencer and envelope generation described in Section 3.7. The "sequence" table, with nine elements, holds a sequence of frequencies. The **phasor~** object at top cycles through the sequence table at 0.6 Hertz. Non-interpolating table lookup (**tabread~** instead of **tabread4~**) is used to read the frequencies in discrete steps. (Such situations, in which we prefer non-interpolating table lookup, are rare.)

The **wrap~** object converts the amplitude-9 sawtooth to a unit-amplitude one as described earlier in Figure 3.8, which is then used to obtain an envelope function from a second wavetable. This is used to control grain size in a looping sampler (from Section 2.6). Here the wavetable consists of six periods of a sinusoid. The grains are smoothed by multiplying by a raised cosine function (**cos~** and **+ 1**).

Example C09.sample.hold.pd (not pictured here) shows a sample-and-hold unit, another useful device for doing control tasks in the audio signal domain.

MIDI-style synthesizer

Example C10.monophonic.synth.pd (Figure 3.16) also implements a monophonic, note-oriented synthesizer, but in this case oriented toward MIDI controllability. Here the tasks of envelope generation and sequencing pitches are handled using control streams instead of audio signals. New control objects are needed for this example:

notein : MIDI note input. Three outlets give the pitch, velocity, and channel of incoming MIDI note-on and note-off events (with note-off events appearing as velocity-zero note-on events). The outputs appear in Pd's customary right-to-left order.

stripnote : filter out note-off messages. This passes (pitch, velocity) pairs through whenever the velocity is nonzero, dropping the others. Unlike

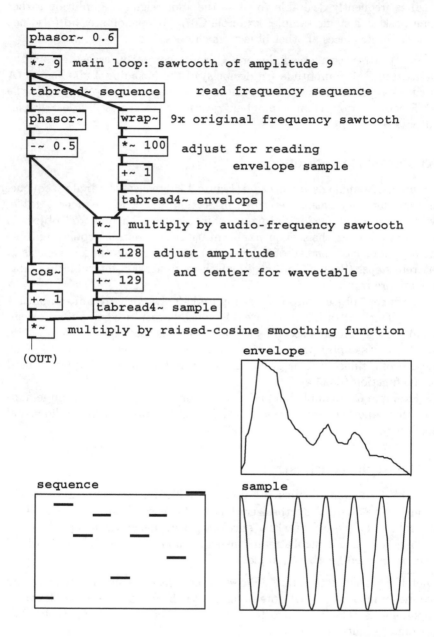

Figure 3.15: An analog-synthesizer-style sequencer.

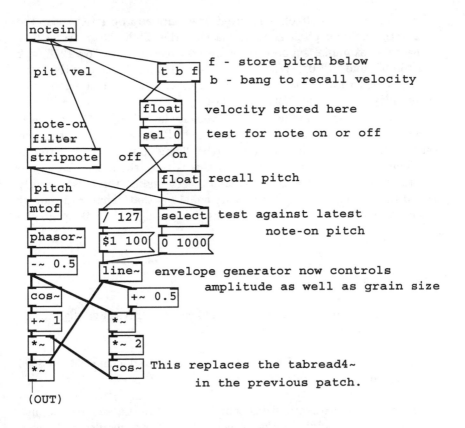

Figure 3.16: A MIDI-style monophonic synthesizer.

notein, stripnote does not directly use hardware MIDI input or output.

| trigger |, | t |: copy a message to outlets in right to left order, with type conversion. The creation arguments ("b" and "f" in this example) specify two outlets, one giving "bang" messages, the other "float" (i.e., numbers). One outlet is created for each creation argument. The outputs appear in Pd's standard right-to-left order.

The patch's control objects feed frequencies to the phasor~ object whenever a MIDI note-on message is received. Controlling the amplitude (via the line~ object) is more difficult. When a note-on message is received, the sel 0 object outputs the velocity at right (because the input failed to match 0); this is divided by the maximum MIDI velocity of 127 and packed into a message for line~ with a time of 100 msec.

However, when a note-off is received, it is only appropriate to stop the sound if the note-off pitch actually matches the pitch the instrument is playing. For example, suppose the messages received are "60 127", "72 127", "60 0", and "72 0". When the note-on at pitch 72 arrives the pitch should change to 72, and then the "60 0" message should be ignored, with the note playing until the "72 0" message.

To accomplish this, first we store the velocity in the upper float object. Second, when the pitch arrives, it too is stored (the lower float object) and then the velocity is tested against zero (the "bang" outlet of t b f recalls the velocity which is sent to sel 0). If this is zero, the second step is to recall the pitch and test it (the select object) against the most recently received note-on pitch. Only if these are equal (so that "bang" appears at the left-hand-side outlet of select) does the message "0 1000" go to the line~ object.

Exercises

1. How many partials of a tone at A 440 can be represented digitally at a sample rate of 44100 Hertz?

2. What frequency would you hear if you synthesized a sinusoid at 88000 Hertz at a sample rate of 44100?

3. Suppose you are synthesizing sound at 44100 Hertz, and are computing 64-sample audio blocks. A control event is scheduled to happen at an elapsed time of exactly one second, using the fast-as-possible update scheme. At what sample does the update actually occur?

4. Sampling at 44100, we wish to approximately play a tone at middle C by repeating a fixed waveform every N samples. What value of N should we choose, and how many cents (Page 7) are we off from the "true" middle C?

5. Two sawtooth waves, of unit amplitude, have frequencies 200 and 300 Hertz, respectively. What is the periodicity of the sum of the two? What if you then wrapped the sum back to the range from 0 to 1? Does this result change depending on the relative phase of the two?

6. Two sawtooth waves, of equal frequency and amplitude and one half cycle out of phase, are summed. What is the waveform of the sum, and what are its amplitude and frequency?

7. What is the relative level, in decibels, of a sawtooth wave's third harmonic (three times the fundamental) compared to that of the fundamental?

8. Suppose you synthesize a 44000-Hertz sawtooth wave at a sample rate of 44100 Hertz. What is the resulting waveform?

9. Using the techniques of Section 3.7, draw a block diagram for generating two phase locked sinusoids at 500 and 700 Hertz.

10. Draw a block diagram showing how to use thresholding to detect when one audio signal exceeds another one in value. (You might want to do this to detect and filter out feedback from speakers to microphones.)

Chapter 4

Automation and Voice Management

It is often desirable to control musical objects or events as aggregates rather than individually. Aggregates might take the form of a series of events spaced in time, in which the details of the events follow from the larger arc (for instance, notes in a melody). Or the individuals might sound simultaneously, as with voices in a chord, or partials in a complex tone. Often some or all properties of the individual elements are best inferred from those of the whole.

A rich collection of tools and ideas has arisen in the electronic music repertory for describing individual behaviors from aggregate ones. In this chapter we cover two general classes of such tools: envelope generators and voice banks. The envelope generator automates behavior over time, and the voice bank over aggregates of simultaneous processes (such as signal generators).

4.1 Envelope Generators

An *envelope generator* (sometimes, and more justly, called a *transient generator*) makes an audio signal that smoothly rises and falls as if to control the loudness of a musical note. Envelope generators were touched on earlier in Section 1.5. Amplitude control by multiplication (Figure 1.4) is the most direct, ordinary way to use one, but there are many other possible uses.

Envelope generators have come in many forms over the years, but the simplest and the perennial favorite is the *ADSR* envelope generator. "ADSR" is an acronym for "Attack, Decay, Sustain, Release", the four segments of

89

Figure 4.1: ADSR envelope as a block diagram, showing the trigger input (a control stream) and the audio output.

the ADSR generator's output. The ADSR generator is turned on and off by a control stream called a "trigger". Triggering the ADSR generator "on" sets off its attack, decay, and sustain segments. Triggering it "off" starts the release segment. Figure 4.1 shows the block diagram representation of an ADSR envelope generator.

There are five parameters controlling the ADSR generator. First, a *level* parameter sets the output value at the end of the attack segment (normally the highest value output by the ADSR generator). Second and third, the *attack* and *decay* parameters give the time duration of the attack and decay segments. Fourth, a *sustain* parameter gives the level of the sustain segment, as a fraction of the level parameter. Finally, the *release* parameter gives the duration of the release segment. These five values, together with the timing of the "on" and "off" triggers, fully determines the output of the ADSR generator. For example, the duration of the sustain portion is equal to the time between "on" and "off" triggers, minus the durations of the attack and decay segments.

Figure 4.2 graphs some possible outputs of an ADSR envelope generator. In part (a) we assume that the "on" and "off" triggers are widely enough separated that the sustain segment is reached before the "off" trigger is received. Parts (b) and (c) of Figure 4.2 show the result of following an "on" trigger quickly by an "off" one: (b) during the decay segment, and (c) even earlier, during the attack. The ADSR generator reacts to these situations by canceling whatever remains of the attack and decay segments and continuing straight to the release segment. Also, an ADSR generator may be retriggered "on" before the release segment is finished or even during the attack, decay, or sustain segments. Part (d) of the figure shows a reattack during the sustain segment, and part (e), during the decay segment.

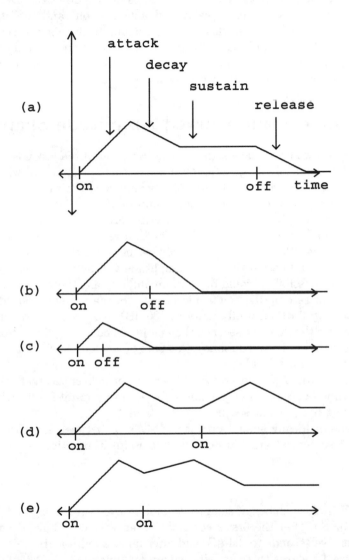

Figure 4.2: ADSR envelope output: (a) with "on" and "off" triggers separated; (b), (c) with early "off" trigger; (d), (e) re-attacked.

The classic application of an ADSR envelope is using a voltage-control keyboard or sequencer to make musical notes on a synthesizer. Depressing and releasing a key (for example) would generate "on" and "off" triggers. The ADSR generator could then control the amplitude of synthesis so that "notes" would start and stop with the keys. In addition to amplitude, the ADSR generator can (and often is) used to control timbre, which can then be made to evolve naturally over the course of each note.

4.2   Linear and Curved Amplitude Shapes

Suppose you wish to fade a signal in over a period of ten seconds—that is, you wish to multiply it by an amplitude-controlling signal $y[n]$ which rises from 0 to 1 in value over $10R$ samples, where R is the sample rate. The most obvious choice would be a linear ramp: $y[n] = n/(10R)$. But this will not turn out to yield a smooth increase in perceived loudness. Over the first second $y[n]$ rises from $-\infty$ dB to -20 dB, over the next four by another 14 dB, and over the remaining five, only by the remaining 6 dB. Over most of the ten second period the rise in amplitude will be barely perceptible.

Another possibility would be to ramp $y[n]$ exponentially, so that it rises at a constant rate in dB. You would have to fix the initial amplitude to be inaudible, say 0 dB (if we fix unity at 100 dB). Now we have the opposite problem: for the first five seconds the amplitude control will rise from 0 dB (inaudible) to 50 dB (pianissimo); this part of the fade-in should only have taken up the first second or so.

A more natural progression would perhaps have been to regard the fade-in as a timed succession of dynamics, 0-ppp-pp-p-mp-mf-f-ff-fff, with each step taking roughly one second.

A fade-in ideally should obey some scale in between logarithmic and linear. A somewhat arbitrary choice, but useful in practice, is the quartic curve:

$$y[n] = \left(\frac{n}{N}\right)^4$$

where N is the number of samples to fade in over (in the example above, it's $10R$). So, after the first second of the ten we would have risen to -80 dB, after five seconds to -24 dB, and after nine, about -4 dB.

Figure 4.3 shows three amplitude transfer functions:

$$f_1(x) = x \quad \text{(linear)}$$

$$f_2(x) = 10^{2(x-1)} \quad \text{(dB to linear)}$$

$$f_3(x) = x^4 \quad \text{(quartic)}$$

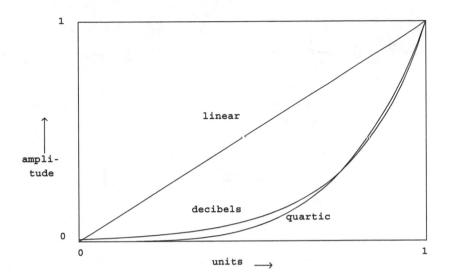

Figure 4.3: Three amplitude transfer functions. The horizontal axis is in linear, logarithmic, or fourth-root units depending on the curve.

The second function converts from dB to linear, arranged so that the input range, from 0 to 1, corresponds to 40 dB. (This input range of 40 dB corresponds to a reasonable dynamic range, allowing 5 dB for each of 8 steps in dynamic.) The quartic curve imitates the exponential (dB) curve fairly well for higher amplitudes, but drops off more rapidly for small amplitudes, reaching true zero at right (whereas the exponential curve only goes down to 1/100).

We can think of the three curves as showing transfer functions, from an abstract control (ranging from 0 to 1) to a linear amplitude. After we choose a suitable transfer function f, we can compute a corresponding amplitude control signal; if we wish to ramp over N samples from silence to unity gain, the control signal would be:

$$y[n] = f(n/N)$$

A block diagram for this is shown in Figure 4.4. Here we are introducing a new type of block to represent the application of a *transfer function*. For now we won't worry about its implementation; depending on the function desired, this might be best done arithmetically or using table lookup.

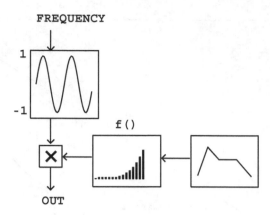

Figure 4.4: Using a transfer function to alter the shape of amplitude curves.

4.3 Continuous and Discontinuous Control Changes

Synthesis algorithms vary widely in their ability to deal with discontinuously changing controls. Until now in this chapter we have assumed that controls must change continuously, and the ADSR envelope generator turns out to be ideally suited for such controls. It may even happen that the maximum amplitude of a note is less than the current value of the amplitude of its predecessor (using the same generator) and the ADSR envelope will simply ramp down (instead of up) to the new value for an attack.

This isn't necessarily desirable, however, in situations where an envelope generator is in charge of some aspect of timbre: perhaps, for example, we don't want the sharpness of a note to decrease during the attack to a milder one, but rather to jump to a much lower value so as always to be able to rise during the attack.

This situation also can arise with pitch envelopes: it may be desirable to slide pitch from one note to the next, or it may be desirable that the pitch trajectory of each note start anew at a point independent of the previous sound.

Two situations arise when we wish to make discontinuous changes to synthesis parameters: either we can simply make them without disruption (for instance, making a discontinuous change in pitch); or else we can't, such as a change in a wavetable index (which makes a discontinuous change in the output). There are even parameters that can't *possibly* be changed continuously; for example, a selection among a collection of wavetables. In general, discontinuously changing the phase of an oscillator or the amplitude of a signal will cause an audible artifact, but phase increments (such as pitches) may jump without bad results.

In those cases where a parameter change can't be made continuously for one reason or another, there are at least two strategies for making the change cleanly: *muting* and *switch-and-ramp*.

4.3.1 Muting

The *muting* technique is to apply an envelope to the output amplitude, which is quickly ramped to zero before the parameter change and then restored afterward. It may or may not be the case that the discontinuous changes will result in a signal that rises smoothly from zero afterward. In Figure 4.5 (part a), we take the example of an amplitude envelope (the output signal of an ADSR generator), and assume that the discontinuous change is to start a new note at amplitude zero.

To change the ADSR generator's output discontinuously we *reset* it. This is a different operation from triggering it; the result is to make it jump to a new value, after which we may either simply leave it there or trigger it anew. Figure 4.5 (part a) shows the effect of resetting and retriggering an ADSR generator.

Below the ADSR generator output we see the appropriate muting signal, which ramps to zero to prepare for the discontinuity. The amount of time we allow for muting should be small (so as to disrupt the previous sound as little as possible) but not so small as to cause audible artifacts in the output. A working example of this type of muting was already shown on Page 81; there we allowed 5 msec for ramping down. The muting signal is multiplied by the output of the process to be de-clicked.

Figure 4.5 (part b) shows the situation in which we suppose the discontinuous change is between two possibly nonzero values. Here the muting signal must not only ramp down as before (in advance of the discontinuity) but must also ramp back up afterward. The ramp-down time need not equal the ramp-up time; these must be chosen, as always, by listening to the output sound.

In general, muting presents the difficulty that you must start the muting operation in advance of making the desired control change. In real-time settings, this often requires that we intentionally delay the control change. This is another reason for keeping the muting time as low as possible. (Moreover, it's a bad idea to try to minimize delay by conditionally omitting the ramp-down period when it isn't needed; a constant delay is much better than one that varies, even if it is smaller on average.)

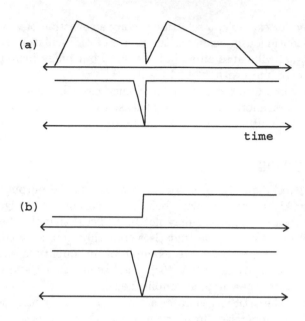

Figure 4.5: Muting technique for hiding discontinuous changes: (a) the envelope (upper graph) is set discontinuously to zero and the muting signal (lower graph) ramps down in advance to prepare for the change, and then is restored (discontinuously) to its previous value; (b) the envelope changes discontinuously between two nonzero values; the muting signal must both ramp down beforehand and ramp back up afterward.

4.3.2   Switch-and-ramp

The *switch-and-ramp* technique also seeks to remove discontinuities resulting from discontinuous control changes, but does so in a different way: by synthesizing an opposing discontinuity which we add to cancel the original one out. Figure 4.6 shows an example in which a synthetic percussive sound (an enveloped sinusoid) starts a note in the middle of a previous one. The attack of the sound derives not from the amplitude envelope but from the initial phase of the sinusoid, as is often appropriate for percussive sounds. The lower graph in the figure shows a compensating audio signal with an opposing discontinuity, which can be added to the upper one to remove the discontinuity. The advantages of this technique over muting are, first, that there need be no delay between the decision to make an attack and the sound of the attack; and second, that any artifacts arising from this technique are more likely to be masked by the new sound's onset.

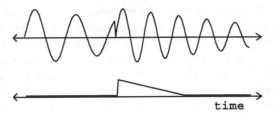

Figure 4.6: The switch-and-ramp technique for canceling out discontinuous changes. A discontinuity (upper graph) is measured and canceled out with a signal having the opposite discontinuity (lower graph), which then decays smoothly.

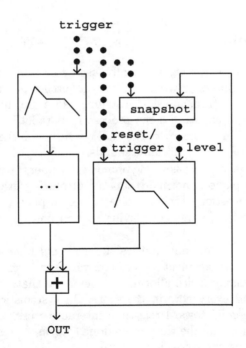

Figure 4.7: Block diagram for the switch-and-ramp technique.

Figure 4.7 shows how the switch-and-ramp technique can be realized in a block diagram. The box marked with ellipsis ("...") may hold any synthesis algorithm, which we wish to interrupt discontinuously so that it restarts from zero (as in, for example, part (a) of the previous figure). At the same time that we trigger whatever control changes are necessary (exemplified by the top ADSR generator), we also reset and trigger another ADSR generator (middle right) to cancel out the discontinuity. The discontinuity is minus the last value of the synthesis output just before it is reset to zero.

To do this we measure the level the ADSR generator must now jump to. This is its own current level (which might not be zero) minus the discontinuity (or equivalently, *plus* the synthesis output's last value). The two are added (by the +~ object at bottom), and then a snapshot is taken. The cancelling envelope generator (at right) is reset discontinuously to this new value, and then triggered to ramp back to zero. The +~ object's output (the sum of the synthesizer output and the discontinuity-cancelling signal) is the de-clicked signal.

4.4   Polyphony

In music, the term *polyphony* is usually used to mean "more than one separate voices singing or playing at different pitches one from another". When speaking of electronic musical instruments we use the term to mean "maintaining several copies of some process in parallel." We usually call each copy a "voice" in keeping with the analogy, although the voices needn't be playing separately distinguishable sounds.

In this language, a piano is a polyphonic instrument, with 88 "voices". Each voice of the piano is normally capable of playing exactly one pitch. There is never a question of which voice to use to play a note of a given pitch, and no question, either, of playing several notes simultaneously at the same pitch.

Many polyphonic electronic musical instruments take a more flexible approach to voice management. Most software synthesis programs (like Csound) use a dynamic voice allocation scheme, so that, in effect, a new voice is created for every note in the score. In systems such as Max or Pd, which are oriented toward real-time interactive use, a *voice bank* is allocated in advance, and the work to be done (playing notes, or whatever) is distributed among the voices in the bank.

This is diagrammed in Figure 4.8, where the several voices each produce one output signal, which are all added to make the total output of the voice bank. Frequently the individual voices will need several separate outputs; for instance, they might output several channels so that they may be panned

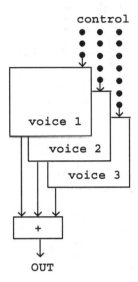

Figure 4.8: A voice bank for polyphonic synthesis.

individually; or they might have individual effect sends so that each may have its own send level.

4.5 Voice Allocation

It is frequently desirable to automate the selection of voices to associate with individual *tasks* (such as notes to play). For example, a musician playing at a keyboard can't practically choose which voice should go with each note played. To automate voice selection we need a voice allocation algorithm, to be used as shown in Figure 4.9.

Armed with a suitable voice allocation algorithm, the control source need not concern itself with the detail of which voice is taking care of which task; algorithmic note generators and sequencers frequently rely on this. On the other hand, musical writing for ensembles frequently specifies explicitly which instrument plays which note, so that the notes will connect to each other end-to-end in a desirable way.

One simple voice allocation algorithm works as shown in Figure 4.10. Here we suppose that the voice bank has only two voices, and we try to allocate voices for the tasks *a*, *b*, *c*, and *d*. Things go smoothly until task *d* comes along, but then we see no free voices (they are taken up by *b* and *c*). We could now elect either to drop task *d*, or else to steal the voice of

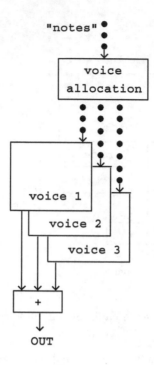

Figure 4.9: Polyphonic voice allocation.

either task b or c. In practice the best choice is usually to steal one. In this particular example, we chose to steal the voice of the oldest task, b.

If we happen to know the length of the tasks b and c at the outset of task d, we may be able to make a better choice of which voice to steal. In this example it might have been better to steal from c, so that d and b would be playing together at the end and not d alone. In some situations this information will be available when the choice must be made, and in some (live keyboard input, for example) it will not.

4.6   Voice Tags

Suppose now that we're using a voice bank to play notes, as in the example above, but suppose the notes a, b, c, and d all have the same pitch, and furthermore that all their other parameters are identical. How can we design a control stream so that, when any one note is turned off, we know which one it is?

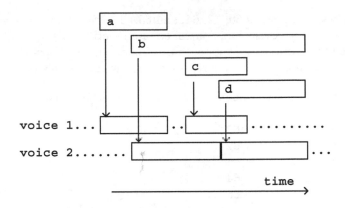

Figure 4.10: A polyphonic voice allocation algorithm, showing voice stealing.

This question doesn't come up if the control source is a clavier keyboard because it's impossible to play more than one simultaneous note on a single key. But it could easily arise algorithmically, or simply as a result of merging two keyboard streams together. Moreover, turning notes off is only the simplest example of a more general problem, which is how, once having set a task off in a voice bank, we can get back to the same voice to guide its evolution as a function of real-time inputs or any other unpredictable factor.

To deal with situations like this we can add one or more *tags* to the message starting a note (or, in general, a task). A tag is any collection of data with which we can later identify the task, which we can then use to search for the voice that is allocated for it.

Taking again the example of Figure 4.10, here is one way we might write those four tasks as a control stream:

start-time end-time   pitch   ...

      1   3   60   ...
      2   8   62
      4   6   64
      5   8   65

In this representation we have no need of tags because each message (each line of text) contains all the information we need in order to specify the entire task. (Here we have assumed that the tasks $a, \ldots, d$ are in fact musical notes with pitches 60, 62, 64, and 65.) In effect we're representing each task as a single event in a control stream (Section 3.3).

On the other hand, if we suppose now that we do not know in advance the length of each note, a better representation would be this one:

time	tag	action	parameters
1	a	start	60 ...
2	b	start	62 ...
3	a	end	
4	c	start	64 ...
5	d	start	65 ...
6	c	end	
8	b	end	
8	d	end	

Here each note has been split into two separate events to start and end it. The labels *a*, ..., *d* are used as tags; we know which start goes with which end since their tags are the same. Note that the tag is not necessarily related at all to the voice that will be used to play each note.

The MIDI standard does not supply tags; in normal use, the pitch of a note serves also as its tag (so tags are constantly being re-used.) If two notes having the same pitch must be addressed separately (to slide their pitches in different directions for example), the MIDI channel may be used (in addition to the note) as a tag.

In real-time music software it is often necessary to pass back and forth between the event-per-task representation and the tagged one above, since the first representation is better suited to storage and graphical editing, while the second is often better suited to real-time operations.

4.7 Encapsulation in Pd

The examples for this chapter will use Pd's *encapsulation* mechanism, which permits the building of patches that may be reused any number of times. One or more object boxes in a Pd patch may be *subpatches*, which are separate patches encapsulated inside the boxes. These come in two types: *one-off subpatches* and *abstractions*. In either case the subpatch appears as an object box in another patch, called the *parent*.

If you type "pd" or "pd my-name" into an object box, this creates a one-off subpatch. The contents of the subpatch are saved as part of the parent patch, in one file. If you make several copies of a subpatch you may change them individually. On the other hand, you can invoke an abstraction by

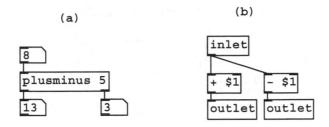

Figure 4.11: Pd's abstraction mechanism: (a) invoking the abstraction, plusminus with 5 as a creation argument; (b) the contents of the file, "plusminus.pd".

typing into the box the name of a Pd patch saved to a file (without the ".pd" extension). In this situation Pd will read that file into the subpatch. In this way, changes to the file propagate everywhere the abstraction is invoked.

A subpatch (either one-off or abstraction) may have inlets and outlets that appear on the box in the parent patch. This is specified using the following objects:

inlet , inlet ~ : create inlets for the object box containing the subpatch. The inlet~ version creates an inlet for audio signals, whereas inlet creates one for control streams. In either case, whatever comes to the inlet of the box in the parent patch comes out of the inlet or inlet~ object in the subpatch.

outlet , outlet ~ : Corresponding objects for output from subpatches.

Pd provides an argument-passing mechanism so that you can parametrize different invocations of an abstraction. If in an object box you type "$1", it is expanded to mean "the first creation argument in my box on the parent patch", and similarly for "$2" and so on. The text in an object box is interpreted at the time the box is created, unlike the text in a message box. In message boxes, the same "$1" means "the first argument of the message I just received" and is interpreted whenever a new message comes in.

An example of an abstraction, using inlets, outlets, and parametrization, is shown in Figure 4.11. In part (a), a patch invokes plusminus in an object box, with a creation argument equal to 5. The number 8 is fed to the plusminus object, and out comes the sum and difference of 8 and 5.

The plusminus object is not defined by Pd, but is rather defined by the patch residing in the file named "plusminus.pd". This patch is shown in part (b) of the figure. The one inlet and two outlet objects correspond

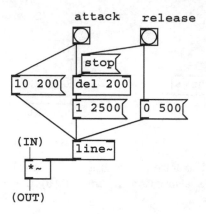

Figure 4.12: Using a `line~` object to generate an ADSR envelope.

to the inlets and outlets of the **plusminus** object. The two "$1" arguments
(to the + and - objects) are replaced by 5 (the creation argument of the
plusminus object).

We have already seen one abstraction in the examples: the **output~**
object introduced in Figure 1.10 (Page 17). That example also shows that
an abstraction may display controls as part of its box on the parent patch;
see the Pd documentation for a description of this feature.

4.8   Examples

ADSR envelope generator

Example D01.envelope.gen.pd (Figure 4.12) shows how the `line~` object
may be used to generate an ADSR envelope to control a synthesis patch
(only the ADSR envelope is shown in the figure). The "attack" button,
when pressed, has two effects. The first (leftmost in the figure) is to set the
`line~` object on its attack segment, with a target of 10 (the peak amplitude)
over 200 msec (the attack time). Second, the attack button sets a **delay**
200 object, so that after the attack segment is done, the decay segment
can start. The decay segment falls to a target of 1 (the sustain level) after
another 2500 msec (the decay time).

The "release" button sends the same `line~` object back to zero over 500
more milliseconds (the release time). Also, in case the **delay** 200 object
happens to be set at the moment the "release" button is pressed, a "stop"

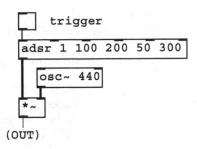

Figure 4.13: Invoking the `adsr` abstraction.

message is sent to it. This prevents the ADSR generator from launching its decay segment after launching its release segment.

In Example D02.adsr.pd (Figure 4.13) we encapsulate the ADSR generator in a Pd abstraction (named `adsr`) so that it can easily be replicated. The design of the `adsr` abstraction makes it possible to control the five ADSR parameters either by supplying creation arguments or by connecting control streams to its inlets.

In this example the five creation arguments (1, 100, 200, 50, and 300) specify the peak level, attack time, decay time, sustain level (as a percentage of peak level), and release time. There are six control inlets: the first to trigger the ADSR generator, and the others to update the values of the five parameters. The output of the abstraction is an audio signal.

This abstraction is realized as shown in Figure 4.14. (You can open this subpatch by clicking on the `adsr` object in the patch.) The only signal objects are `line~` and `outlet~`. The three `pack` objects correspond to the three message objects from the earlier Figure 4.12. From left to right, they take care of the attack, decay, and release segments.

The attack segment goes to a target specified as "$1" (the first creation argument of the abstraction) over "$2" milliseconds; these values may be overwritten by sending numbers to the "peak level" and "attack" inlets. The release segment is similar, but simpler, since the target is always zero. The hard part is the decay segment, which again must be set off after a delay equal to the attack time (the `del $2` object). The sustain level is calculated from the peak level and the sustain percentage (multiplying the two and dividing by 100).

The trigger inlet, if sent a number other than zero, triggers an onset (the attack and decay segments), and if sent zero, triggers the release segment. Furthermore, the ADSR generator may be reset to zero by sending a negative trigger (which also generates an onset).

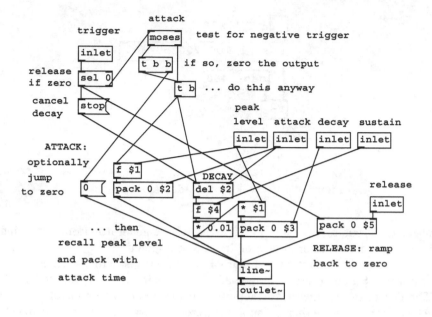

Figure 4.14: Inside the adsr abstraction.

Transfer functions for amplitude control

Section 4.2 described using ADSR envelopes to control amplitude, for which exponential or quartic-curve segments often give more natural-sounding results than linear ones. Patches D03.envelope.dB.pd and D04.envelope.quartic.pd (the latter shown in Figure 4.15) demonstrate the use of decibel and quartic segments. In addition to amplitude, in Example D04.envelope.quartic.pd the frequency of a sound is also controlled, using linear and quartic shapes, for comparison.

Since converting decibels to linear amplitude units is a costly operation (at least when compared to an oscillator or a ramp generator), Example D03.envelope.dB.pd uses table lookup to implement the necessary transfer function. This has the advantage of efficiency, but the disadvantage that we must decide on the range of admissible values in advance (here from 0 to 120 dB).

For a quartic segment as in Example D04.envelope.quartic.pd no table lookup is required; we simply square the line~ object's output signal twice in succession. To compensate for raising the output to the fourth power, the target values sent to the line~ object must be the fourth root of the

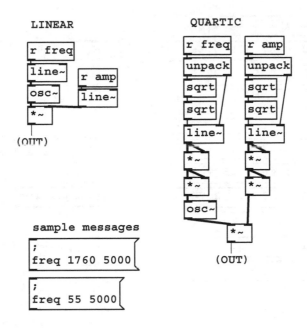

Figure 4.15: Linear and quartic transfer functions for changing amplitude and pitch.

desired ones. Thus, messages to ramp the frequency or amplitude are first unpacked to separate the target and time interval, the target's fourth root is taken (via two square roots in succession), and the two are then sent to the line~ object. Here we have made use of one new Pd object:

unpack : unpack a list of numbers (and/or symbols) and distribute them to separate outlets. As usual the outputs appear in right-to-left order. The number of outlets and their types are determined by the creation arguments. (See also pack, Page 50).

The next two patches, D05.envelope.pitch.pd and D06.envelope.portamento.dp, use an ADSR envelope generator to make a pitch envelope and a simple line~ object, also controlling pitch, to make portamento. In both cases exponential segments are desirable, and they are calculated using table lookup.

Additive synthesis: Risset's bell

The abstraction mechanism of Pd, which we used above to make a reusable ADSR generator, is also useful for making voice banks. Here we will use

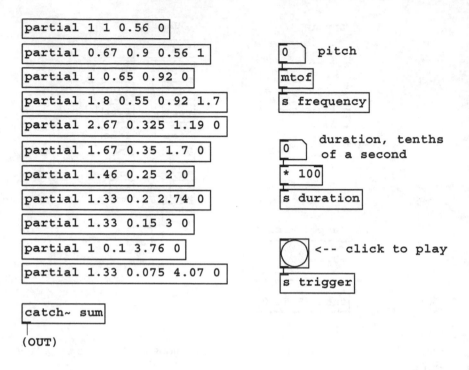

Figure 4.16: A Pd realization of Jean-Claude Risset's bell instrument. The bell sound is made by summing 11 sinusoids, each made by a copy of the **partial** abstraction.

abstractions to organize banks of oscillators for additive synthesis. There are many possible ways of organizing oscillator banks besides those shown here.

The simplest and most direct organization of the sinusoids is to form partials to add up to a note. The result is monophonic, in the sense that the patch will play only one note at a time, which, however, will consist of several sinusoids whose individual frequencies and amplitudes might depend both on those of the note we're playing, and also on their individual placement in a harmonic (or inharmonic) overtone series.

Example D07.additive.pd (Figure 4.16) uses a bank of 11 copies of an abstraction named **partial** (Figure 4.17) in an imitation of a well-known bell instrument by Jean-Claude Risset. As described in [DJ85, p. 94], the bell sound has 11 partials, each with its own relative amplitude, frequency, and duration.

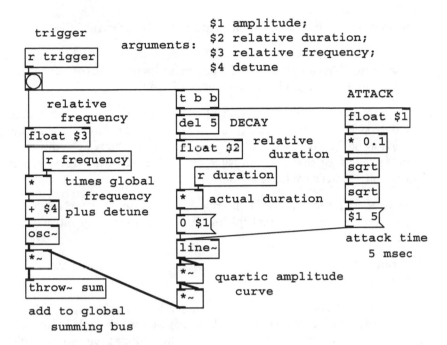

Figure 4.17: The `partial` abstraction used by Risset's bell instrument from Figure 4.16.

For each note, the `partial` abstraction computes a simple (quartic) amplitude envelope consisting only of an attack and a decay segment; there is no sustain or release segment. This is multiplied by a sinusoid, and the product is added into a summing bus. Two new object classes are introduced to implement the summing bus:

`catch~`: define and output a summing bus. The creation argument ("sum-bus" in this example) gives the summing bus a name so that `throw~` objects below can refer to it. You may have as many summing busses (and hence `catch~` objects) as you like but they must all have different names.

`throw~`: add to a summing bus. The creation argument selects which summing bus to use.

The control interface is crude: number boxes control the "fundamental" frequency of the bell and its duration. Sending a "bang" message to the s `trigger` object starts a note. (The note then decays over the period of time controlled by the duration parameter; there is no separate trigger to stop the note). There is no amplitude control except via the `output~` object.

The four arguments to each invocation of the `partial` abstraction specify:

1. amplitude. The amplitude of the partial at its peak, at the end of the attack and the beginning of the decay of the note.

2. relative duration. This is multiplied by the overall note duration (controlled in the main patch) to determine the duration of the decay portion of the sinusoid. Individual partials may thus have different decay times, so that some partials die out faster than others, under the main patch's overall control.

3. relative frequency. As with the relative duration, this controls each partial's frequency as a multiple of the overall frequency controlled in the main patch.

4. detune. A frequency in Hertz to be added to the product of the global frequency and the relative frequency.

Inside the `partial` abstraction, the amplitude is simply taken directly from the "$1" argument (multiplying by 0.1 to adjust for the high individual amplitudes); the duration is calculated from the `r duration` object, multiplying it by the "$2" argument. The frequency is computed as $fp + d$ where f is the global frequency (from the `r frequency` object), p is the relative frequency of the partial, and d is the detune frequency.

Additive synthesis: spectral envelope control

The next patch example, D08.table.spectrum.pd (Figure 4.18), shows a very different application of additive synthesis from the previous one. Here the sinusoids are managed by the `spectrum-partial` abstraction shown in Figure 4.19. Each partial computes its own frequency as in the previous patch. Each partial also computes its own amplitude periodically (when triggered by the `r poll-table` object), using a `tabread4` object. The contents of the table (which has a nominal range of 50 dB) are converted to linear units and used as an amplitude control in the usual way.

A similar example, D09.shepard.tone.pd (not pictured), makes a Shepard tone using the same technique. Here the frequencies of the sinusoids sweep over a fixed range, finally jumping from the end back to the beginning and repeating. The spectral envelope is arranged to have a peak at the middle of the pitch range and drop off to inaudibility at the edges of the range so that we hear only the continuous sweeping and not the jumping. The result is a famous auditory conundrum, an indefinitely ascending or descending tone.

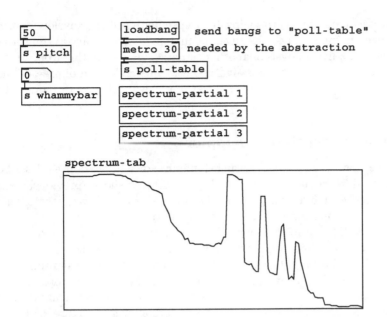

Figure 4.18: Additive synthesis for a specified spectral envelope, drawn in a table.

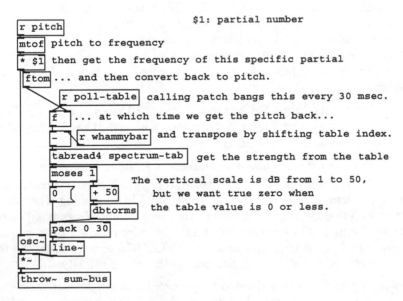

Figure 4.19: The spectrum-partial abstraction used in Figure 4.18.

The technique of synthesizing to a specified spectral envelope can be generalized in many ways: the envelope may be made to vary in time either as a result of a live analysis of another signal, or by calculating from a set of compositional rules, or by cross-fading between a collection of pre-designed spectral envelopes, or by frequency-warping the envelopes, to name a few possibilities.

Polyphonic synthesis: sampler

We move now to an example using dynamic voice allocation as described in Section 4.5. In the additive synthesis examples shown previously, each voice is used for a fixed purpose. In the present example, we allocate voices from a bank as needed to play notes in a control stream.

Example D11.sampler.poly.pd (Figure 4.20) shows the polyphonic sampler, which uses the abstraction `sampvoice` (whose interior is shown in Figure 4.21). Techniques for altering the pitch and other parameters in a one-shot sampler are introduced in Example D10.sampler.notes.pd (not shown) which in turn is derived from the original one-shot sampler from the previous chapter (C05.sampler.oneshot.pd, shown in Figure 3.14).

The `sampvoice` objects in Figure 4.20 are arranged in a different kind of summing bus from the ones before; here, each one adds its own output to the signal on its inlet, and puts the sum on its outlet. At the bottom of the eight objects, the outlet therefore holds the sum of all eight. This has the advantage of being more explicit than the `throw~` / `catch~` busses, and is preferable when visual clutter is not a problem.

The main job of the patch, though, is to distribute the "note" messages to the `sampvoice` objects. To do this we must introduce some new Pd objects:

| mod |: Integer modulus. For instance, 17 mod 10 gives 7, and -2 mod 10 gives 8. There is also an integer division object named `div` ; dividing 17 by 10 via `div` gives 1, and -2 by 10 gives -1.

| poly |: Polyphonic voice allocator. Creation arguments give the number of voices in the bank and a flag (1 if voice stealing is needed, 0 if not). The inlets are a numeric tag at left and a flag at right indicating whether to start or stop a voice with the given tag (nonzero numbers meaning "start" and zero, "stop"). The outputs are, at left, the voice number, the tag again at center, and the start/stop flag at right. In MIDI applications, the tag can be pitch and the start/stop flag can be the note's velocity.

| makenote |: Supply delayed note-off messages to match note-on messages. The inlets are a tag and start/stop flag ("pitch" and "velocity" in MIDI usage) and the desired duration in milliseconds. The tag/flag pair are

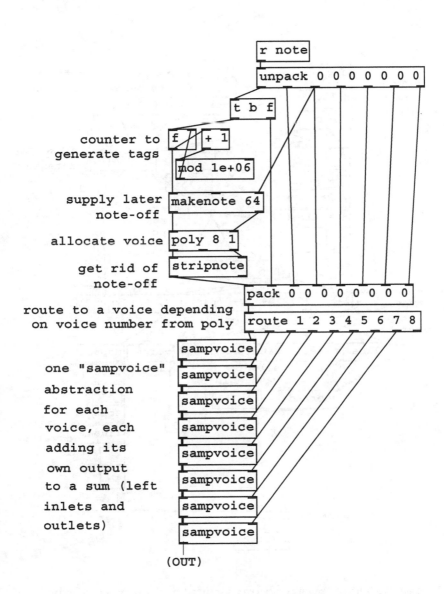

Figure 4.20: A polyphonic sampler demonstrating voice allocation and use of tags.

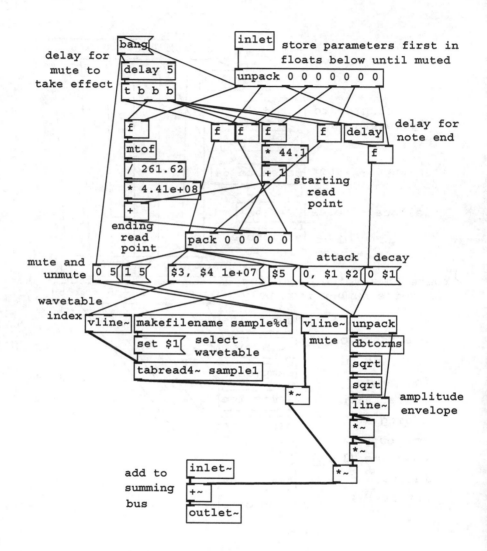

Figure 4.21: The **sampvoice** abstraction used in the polyphonic sampler of Figure 4.20.

repeated to the two outlets as they are received; then, after the delay, the tag is repeated with flag zero to stop the note after the desired duration.

The "note" messages contain fields for pitch, amplitude, duration, sample number, start location in the sample, rise time, and decay time. For instance, the message,

60 90 1000 2 500 10 20

(if received by the r note object) means to play a note at pitch 60 (MIDI units), amplitude 90 dB, one second long, from the wavetable named "sample2", starting at a point 500 msec into the wavetable, with rise and decay times of 10 and 20 msec.

After unpacking the message into its seven components, the patch creates a tag for the note. To do this, first the t b f object outputs a bang after the last of the seven parameters appear separately. The combination of the +, f, and mod objects acts as a counter that repeats after a million steps, essentially generating a unique number corresponding to the note.

The next step is to use the poly object to determine which voice to play which note. The poly object expects separate messages to start and stop tasks (i.e., notes). So the tag and duration are first fed to the makenote object, whose outputs include a flag ("velocity") at right and the tag again at left. For each tag makenote receives, two pairs of numbers are output, one to start the note, and another, after a delay equal to the note duration, to stop it.

Having treated poly to this separated input, we now have to strip the messages corresponding to the ends of notes, since we really only need combined "note" messages with duration fields. The stripnote object does this job. Finally, the voice number we have calculated is prepended to the seven parameters we started with (the pack object), so that the output of the pack object looks like this:

4 60 90 1000 2 500 10 20

where the "4" is the voice number output by the poly object. The voice number is used to route the message to the desired voice using the route object. The appropriate sampvoice object then gets the original list starting with "60".

Inside the sampvoice object (Figure 4.21), the message is used to control the tabread4~ and surrounding line~ and vline~ objects. The control takes place with a delay of 5 msec as in the earlier sampler example. Here, however, we must store the seven parameters of the note (earlier there were no parameters). This is done using the six f objects, plus the right inlet of the rightmost delay object. These values are used after the delay of 5 msec. This is done in tandem with the muting mechanism described on Page 95, using another `vline~` object.

When the 5 msec have elapsed, the `vline~` object in charge of generating the wavetable index gets its marching orders (and, simultaneously, the wavetable number is set for the `tabread4~` object and the amplitude envelope generator starts its attack.) The wavetable index must be set discontinuously to the starting index, then ramped to an ending index over an appropriate time duration to obtain the needed transposition. The starting index in samples is just 44.1 times the starting location in milliseconds, plus one to allow for four-point table interpolation. This becomes the third number in a packed list generated by the `pack` object at the center of the voice patch.

We arbitrarily decide that the ramp will last ten thousand seconds (this is the "1e+07" appearing in the message box sent to the wavetable index generator), hoping that this is at least as long as any note we will play. The ending index is the starting index plus the number of samples to ramp through. At a transposition factor of one, we should move by 441,000,000 samples during those 10,000,000 milliseconds, or proportionally more or less depending on the transposition factor. This transposition factor is computed by the `mtof` object, dividing by 261.62 (the frequency corresponding to MIDI note 60) so that a specified "pitch" of 60 results in a transposition factor of one.

These and other parameters are combined in one message via the `pack` object so that the following message boxes can generate the needed control messages. The only novelty is the `makefilename` object, which converts numbers such as "2" to symbols such as "sample2" so that the `tabread4~` object's wavetable may be set.

At the bottom of the voice patch we see how a summing bus is implemented inside a subpatch; an `inlet~` object picks up the sum of all the preceding voices, the output of the current voice is added in, and the result is sent on to the next voice via the `outlet~` object.

Exercises

1. What input to a fourth-power transfer function gives an output of -12 dB, if an input of 1 outputs 0 dB?

2. An envelope generator rises from zero to a peak value during its attack segment. How many decibels less than the peak has the output reached halfway into the attack, assuming linear output? Fourth-power output?

3. What power-law transfer function (i.e. what choice of n for the function $f(x) = x^n$) would you use if you wish the halfway-point value to be -12 decibels?

4. Suppose you wish to cross-fade two signals, i.e., to ramp one signal in and simultaneously another one out. If they have equal power and are uncorrelated, a linear cross-fade would result in a drop of 3 decibels halfway though the cross-fade. What power law would you use to maintain constant power throughout the cross-fade?

5. A three-note chord, lasting 1.5 seconds, is played starting once every second. How many voices would be needed to synthesize this without cutting off any notes?

6. Suppose a synthesis patch gives output between -1 and 1. While a note is playing, a new note is started using the "rampdown" voice-stealing technique. What is the maximum output?

Chapter 5

Modulation

Having taken a two-chapter detour into aspects of control and organization in electronic music, we return to describing audio synthesis and processing techniques. So far we have seen additive and wavetable-based methods. In this chapter we will introduce three so-called *modulation* techniques: *amplitude modulation*, *frequency modulation*, and *waveshaping*. The term "modulation" refers loosely to any technique that systematically alters the shape of a waveform by bending its graph vertically or horizontally. Modulation is widely used for building synthetic sounds with various families of *spectra*, for which we must develop some terminology before getting into the techniques.

5.1 Taxonomy of Spectra

Figure 5.1 introduces a way of visualizing the *spectrum* of an audio signal. The spectrum describes, roughly speaking, how the signal's power is distributed into frequencies. (Much more precise definitions can be given than those that we'll develop here, but they would require more mathematical background.)

Part (a) of the figure shows the spectrum of a *harmonic signal*, which is a periodic signal whose fundamental frequency is in the range of perceptible pitches, roughly between 50 and 4000 Hertz. The Fourier series (Page 13) gives a description of a periodic signal as a sum of sinusoids. The frequencies of the sinusoids are in the ratio $0 : 1 : 2 : \cdots$. (The constant term in the Fourier series may be thought of as a sinusoid,

$$a_0 = a_0 \cos(0 \cdot \omega n)$$

whose frequency is zero.)

119

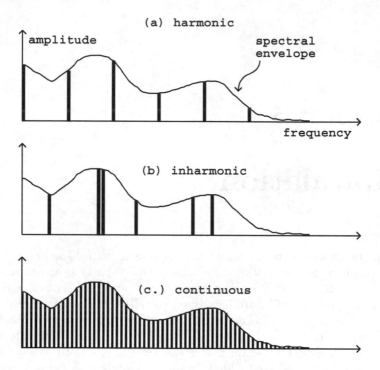

Figure 5.1: A taxonomy of timbres. The spectral envelope describes the shape of the spectrum. The sound may be discretely or continuously distributed in frequency; if discretely, it may be harmonic or inharmonic.

In a harmonic signal, the power shown in the spectrum is concentrated on a discrete subset of the frequency axis (a discrete set consists of isolated points, only finitely many in any bounded interval). We call this a *discrete* spectrum. Furthermore, the frequencies where the signal's power lies are in the $0 : 1 : 2 \cdots$ ratio that arises from a periodic signal. (It's not necessary for *all* of the harmonic frequencies to be present; some harmonics may have zero amplitude.) For a harmonic signal, the graph of the spectrum shows the amplitudes of the partials of the signals. Knowing the amplitudes and phases of all the partials fully determines the original signal.

Part (b) of the figure shows a spectrum which is also discrete, so that the signal can again be considered as a sum of a series of partials. In this case, however, there is no fundamental frequency, i.e., no audible common submultiple of all the partials. This is called an *inharmonic* signal. (The terms *harmonic* and *inharmonic* may be used to describe both the signals and their spectra.)

When dealing with discrete spectra, we report a partial's amplitude in a slightly non-intuitive way. Each component sinusoid,

$$a\cos(\omega n + \phi)$$

only counts as having amplitude $a/2$ as long as the angular frequency ω is nonzero. But for a component of zero frequency, for which $\omega = \phi = 0$, the amplitude is given as a—without dividing by two. (Components of zero frequency are often called *DC* components; "DC" is historically an acronym for "direct current"). These conventions for amplitudes in spectra will simplify the mathematics later in this chapter; a deeper reason for them will become apparent in Chapter 7.

Part (c) of the figure shows a third possibility: the spectrum might not be concentrated into a discrete set of frequencies, but instead might be spread out among all possible frequencies. This can be called a *continuous*, or *noisy* spectrum. Spectra don't necessarily fall into either the discrete or continuous categories; real sounds, in particular, are usually somewhere in between.

Each of the three parts of the figure shows a continuous curve called the *spectral envelope*. In general, sounds don't have a single, well-defined spectral envelope; there may be many ways to draw a smooth-looking curve through a spectrum. On the other hand, a spectral envelope may be specified intentionally; in that case, it is usually clear how to make a spectrum conform to it. For a discrete spectrum, for example, we could simply read off, from the spectral envelope, the desired amplitude of each partial and make it so.

A sound's pitch can sometimes be inferred from its spectrum. For discrete spectra, the pitch is primarily encoded in the frequencies of the partials. Harmonic signals have a pitch determined by their fundamental frequency; for inharmonic ones, the pitch may be clear, ambiguous, or absent altogether, according to complex and incompletely understood rules. A noisy spectrum may also have a perceptible pitch if the spectral envelope contains one or more narrow peaks. In general, a sound's loudness and timbre depend more on its spectral envelope than on the frequencies in the spectrum, although the distinction between continuous and discrete spectra may also be heard as a difference in timbre.

Timbre, as well as pitch, may evolve over the life of a sound. We have been speaking of spectra here as static entities, not considering whether they change in time or not. If a signal's pitch and timbre are changing over time, we can think of the spectrum as a time-varying description of the signal's momentary behavior.

This way of viewing sounds is greatly oversimplified. The true behavior of audible pitch and timbre has many aspects which can't be explained in terms of this model. For instance, the timbral quality called "roughness" is sometimes thought of as being reflected in rapid changes in the spectral envelope over time. The simplified description developed here is useful nonetheless in discussions about how to construct discrete or continuous spectra for a wide variety of musical purposes, as we will begin to show in the rest of this chapter.

5.2   Multiplying Audio Signals

We have been routinely adding audio signals together, and multiplying them by slowly-varying signals (used, for example, as amplitude envelopes) since Chapter 1. For a full understanding of the algebra of audio signals we must also consider the situation where two audio signals, neither of which may be assumed to change slowly, are multiplied. The key to understanding what happens is the Cosine Product Formula:

$$\cos(a)\cos(b) = \frac{1}{2}\Big[\cos(a+b) + \cos(a-b)\Big]$$

To see why this formula holds, we can use the formula for the cosine of a sum of two angles:

$$\cos(a+b) = \cos(a)\cos(b) - \sin(a)\sin(b)$$

to evaluate the right hand side of the cosine product formula; it then simplifies to the left hand side.

We can use this formula to see what happens when we multiply two sinusoids (Page 1):

$$\cos(\alpha n + \phi)\cos(\beta n + \xi)$$

$$= \frac{1}{2}\Big[\cos\left((\alpha+\beta)n + (\phi+\xi)\right) + \cos\left((\alpha-\beta)n + (\phi-\xi)\right)\Big]$$

In words, multiply two sinusoids and you get a result with two partials, one at the sum of the two original frequencies, and one at their difference. (If the difference $\alpha - \beta$ happens to be negative, simply switch the original two sinusoids and the difference will then be positive.) These two new components are called *sidebands*.

This gives us a technique for shifting the component frequencies of a sound, called *ring modulation*, which is shown in its simplest form in Figure 5.2. An oscillator provides a *carrier signal*, which is simply multiplied by the input. In this context the input is called the *modulating signal*. The term "ring modulation" is often used more generally to mean multiplying any two signals together, but here we'll just consider using a

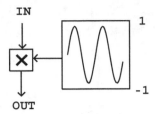

Figure 5.2: Block diagram for ring modulating an input signal with a sinusoid.

sinusoidal carrier signal. (The technique of ring modulation dates from the analog era [Str95]; digital multipliers now replace both the VCA (Section 1.5) and the ring modulator.)

Figure 5.3 shows a variety of results that may be obtained by multiplying a (modulating) sinusoid of angular frequency α and peak amplitude $2a$, by a (carrier) sinusoid of angular frequency β and peak amplitude 1:

$$[2a \cos(\alpha n)] \cdot [\cos(\beta n)]$$

(For simplicity the phase terms are omitted.) Each part of the figure shows both the modulation signal and the result in the same spectrum. The modulating signal appears as a single frequency, α, at amplitude a. The product in general has two component frequencies, each at an amplitude of $a/2$.

Parts (a) and (b) of the figure show "general" cases where α and β are nonzero and different from each other. The component frequencies of the output are $\alpha + \beta$ and $\alpha - \beta$. In part (b), since $\alpha - \beta < 0$, we get a negative frequency component. Since cosine is an even function, we have

$$\cos((\alpha - \beta)n) = \cos((\beta - \alpha)n)$$

so the negative component is exactly equivalent to one at the positive frequency $\beta - \alpha$, at the same amplitude.

In the special case where $\alpha = \beta$, the second (difference) sideband has zero frequency. In this case phase will be significant so we rewrite the product with explicit phases, replacing β by α, to get:

$$2a \cos(\alpha n + \phi) \cos(\alpha n + \xi)$$

$$= a \cos(2\alpha n + (\phi + \xi)) + a \cos(\phi - \xi)$$

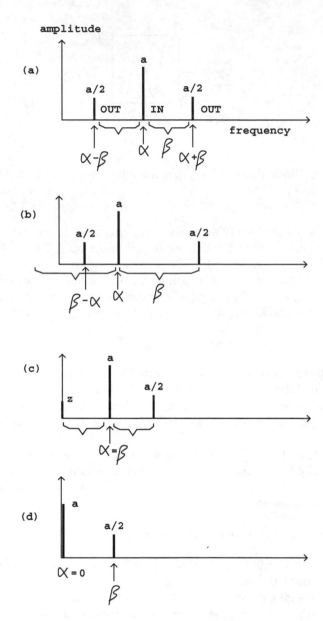

Figure 5.3: Sidebands arising from multiplying two sinusoids of frequency α and β: (a) with $\alpha > \beta > 0$; (b) with $\beta > \alpha$ so that the lower sideband is reflected about the $f = 0$ axis; (c) with $\alpha = \beta$, for which the amplitude of the zero-frequency sideband depends on the phases of the two sinusoids; (d) with $\alpha = 0$.

The second term has zero frequency; its amplitude depends on the relative phase of the two sinusoids and ranges from $+a$ to $-a$ as the phase difference $\phi - \xi$ varies from 0 to π radians. This situation is shown in part (c) of Figure 5.3.

Finally, part (d) shows a carrier signal whose frequency is zero. Its value is the constant a (not $2a$; zero frequency is a special case). Here we get only one sideband, of amplitude $a/2$ as usual.

We can use the distributive rule for multiplication to find out what happens when we multiply signals together which consist of more than one partial each. For example, in the situation above we can replace the signal of frequency α with a sum of several sinusoids, such as:

$$a_1 \cos(\alpha_1 n) + \cdots + a_k \cos(\alpha_k n)$$

Multiplying by the signal of frequency β gives partials at frequencies equal to:

$$\alpha_1 + \beta, \alpha_1 - \beta, \ldots, \alpha_k + \beta, \alpha_k - \beta$$

As before if any frequency is negative we take its absolute value.

Figure 5.4 shows the result of multiplying a complex periodic signal (with several components tuned in the ratio 0:1:2:···) by a sinusoid. Both the spectral envelope and the component frequencies of the result are changed according to relatively simple rules.

The resulting spectrum is essentially the original spectrum combined with its reflection about the vertical axis. This combined spectrum is then shifted to the right by the carrier frequency. Finally, if any components of the shifted spectrum are still left of the vertical axis, they are reflected about it to make positive frequencies again.

In part (b) of the figure, the carrier frequency (the frequency of the sinusoid) is below the fundamental frequency of the complex signal. In this case the shifting is by a relatively small distance, so that re-folding the spectrum at the end almost places the two halves on top of each other. The result is a spectral envelope roughly the same as the original (although half as high) and a spectrum twice as dense.

A special case, not shown, is to use a carrier frequency half the fundamental. In this case, pairs of partials will fall on top of each other, and will have the ratios $1/2 : 3/2 : 5/2 : \cdots$ to give an odd-partial-only signal an octave below the original. This is a very simple and effective octave divider for a harmonic signal, assuming you know or can find its fundamental frequency. If you want even partials as well as odd ones (for the octave-down signal), simply mix the original signal with the modulated one.

Part (c) of the figure shows the effect of using a modulating frequency much higher than the fundamental frequency of the complex signal. Here

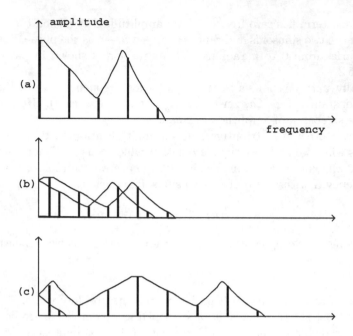

Figure 5.4: Result of ring modulation of a complex signal by a pure sinusoid: (a) the original signal's spectrum and spectral envelope; (b) modulated by a relatively low carrier frequency (1/3 of the fundamental); (c) modulated by a higher frequency, 10/3 of the fundamental.

the unfolding effect is much more clearly visible (only one partial, the leftmost one, had to be reflected to make its frequency positive). The spectral envelope is now widely displaced from the original; this displacement is often a more strongly audible effect than the relocation of partials.

As another special case, the carrier frequency may be a multiple of the fundamental of the complex periodic signal; then the partials all land back on other partials of the same fundamental, and the only effect is the shift in spectral envelope.

5.3   Waveshaping

Another approach to modulating a signal, called *waveshaping*, is simply to pass it through a suitably chosen nonlinear function. A block diagram for doing this is shown in Figure 5.5. The function $f()$ (called the *transfer function*) distorts the incoming waveform into a different shape. The new

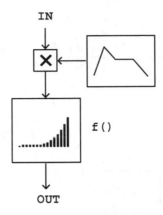

Figure 5.5: Block diagram for waveshaping an input signal using a nonlinear function $f()$. An amplitude adjustment step precedes the function lookup, to take advantage of the different effect of the wavetable lookup at different amplitudes.

shape depends on the shape of the incoming wave, on the transfer function, and also—crucially—on the amplitude of the incoming signal. Since the amplitude of the input waveform affects the shape of the output waveform (and hence the timbre), this gives us an easy way to make a continuously varying family of timbres, simply by varying the input level of the transformation. For this reason, it is customary to include a leading amplitude control as part of the waveshaping operation, as shown in the block diagram.

The amplitude of the incoming waveform is called the waveshaping *index*. In many situations a small index leads to relatively little distortion (so that the output closely resembles the input) and a larger one gives a more distorted, richer timbre.

Figure 5.6 shows a familiar example of waveshaping, in which $f()$ amounts to a *clipping function*. This example shows clearly how the input amplitude—the index—can affect the output waveform. The clipping function passes its input to the output unchanged as long as it stays in the interval between -0.3 and +0.3. So when the input does not exceed 0.3 in absolute value, the output is the same as the input. But when the input grows past the limits, the output stays within; and as the amplitude of the signal increases the effect of this clipping action is progressively more severe. In the figure, the input is a decaying sinusoid. The output evolves

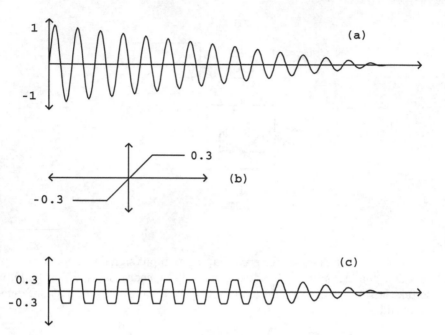

Figure 5.6: Clipping as an example of waveshaping: (a) the input, a decaying sinusoid; (b) the waveshaping function, which clips its input to the interval between -0.3 and +0.3; (c) the result.

from a nearly square waveform at the beginning to a pure sinusoid at the end. This effect will be well known to anyone who has played an instrument through an overdriven amplifier. The louder the input, the more distorted will be the output. For this reason, waveshaping is also sometimes called *distortion*.

Figure 5.7 shows a much simpler and easier to analyse situation, in which the transfer function simply squares the input:

$$f(x) = x^2$$

For a sinusoidal input,

$$x[n] = a \cos(\omega n + \phi)$$

we get

$$f(x[n]) = \frac{a^2}{2} \left(1 + \cos(2\omega n + 2\phi)\right)$$

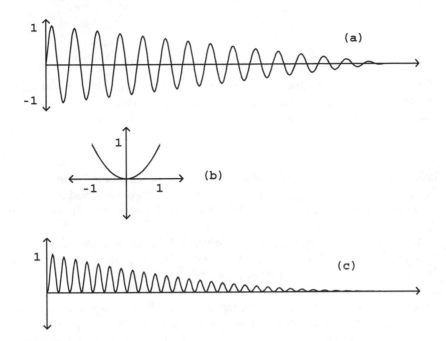

Figure 5.7: Waveshaping using a quadratic transfer function $f(x) = x^2$: (a) the input; (b) the transfer function; (c) the result, sounding at twice the original frequency.

If the amplitude a equals one, this just amounts to ring modulating the sinusoid by a sinusoid of the same frequency, whose result we described in the previous section: the output is a DC (zero-frequency) sinusoid plus a sinusoid at twice the original frequency. However, in this waveshaping example, unlike ring modulation, the amplitude of the output grows as the square of the input.

Keeping the same transfer function, we now consider the effect of sending in a combination of two sinusoids with amplitudes a and b, and angular frequencies α and β. For simplicity, we'll omit the initial phase terms. We set:

$$x[n] = a\cos(\alpha n) + b\cos(\beta n)$$

and plugging this into $f()$ gives

$$f(x[n]) = \frac{a^2}{2}\left(1 + \cos(2\alpha n)\right) +$$

$$+\frac{b^2}{2}\left(1+\cos(2\beta n)\right)$$

$$+ab\left[\cos((\alpha+\beta)n)+\cos((\alpha-\beta)n)\right]$$

The first two terms are just what we would get by sending the two sinusoids through separately. The third term is twice the product of the two input terms, which comes from the middle, cross term in the expansion,

$$f(x+y)=x^2+2xy+y^2$$

This effect, called *intermodulation*, becomes more and more dominant as the number of terms in the input increases; if there are k sinusoids in the input there are only k "straight" terms in the product, but there are $(k^2-k)/2$ intermodulation terms.

In contrast with ring modulation, which is a linear function of its input signal, waveshaping is nonlinear. While we were able to analyze linear processes by considering their action separately on all the components of the input, in this nonlinear case we also have to consider the interactions between components. The results are far more complex—sometimes sonically much richer, but, on the other hand, harder to understand or predict.

In general, we can show that a periodic input, no matter how complex, will repeat at the same period after waveshaping: if the period is τ so that

$$x[n+\tau]=x[n]$$

and temporarily setting the index $a=1$,

$$f(x[n+\tau])=f(x[n])$$

(In some special cases the output can repeat at a submultiple of τ, so that we get a harmonic of the input as a result; this happened for example in Figure 5.4.)

Combinations of periodic tones at consonant intervals can give rise to distortion products at subharmonics. For instance, if two periodic signals x and y are a musical fourth apart (periods in the ratio 4:3), then the sum of the two repeats at the lower rate given by the common subharmonic. In equations we would have:

$$x[t+\tau/3]=x[t]$$

$$y[t+\tau/4]=y[t]$$

which implies

$$x[t+\tau]+y[t+\tau]=x[t]+y[t]$$

and so the distorted sum $f(x+y)$ would repeat after a period of τ:

$$f(x+y)[n+\tau] = f(x+y)[n]$$

This has been experienced by every electric guitarist who has set the amplifier to "overdrive" and played the open B and high E strings together: the distortion product sometimes sounds at the pitch of the low E string, two octaves below the high one.

To get a somewhat more explicit analysis of the effect of waveshaping on an incoming signal, it is sometimes useful to write the function f as a finite or infinite *power series*:

$$f(x) = f_0 + f_1 x + f_2 x^2 + f_3 x^3 + \cdots$$

If the input signal $x[n]$ is a unit-amplitude sinusoid, $\cos(\omega n)$, we can consider the action of the above terms separately:

$$f(a \cdot x[n]) = f_0 + a f_1 \cos(\omega n) + a^2 f_2 \cos^2(\omega n) + a^3 f_3 \cos^3(\omega n) + \cdots$$

Since the terms of the series are successively multiplied by higher powers of the index a, a lower value of a will emphasize the earlier terms more heavily, and a higher value will emphasize the later ones.

The individual terms' spectra can be found by applying the cosine product formula repeatedly:

$$1 = \cos(0)$$

$$x[n] = \cos(\omega n)$$

$$x^2[n] = \frac{1}{2} + \frac{1}{2}\cos(2\omega n)$$

$$x^3[n] = \frac{1}{4}\cos(-\omega n) + \frac{2}{4}\cos(\omega n) + \frac{1}{4}\cos(3\omega n)$$

$$x^4[n] = \frac{1}{8}\cos(-2\omega n) + \frac{3}{8}\cos(0) + \frac{3}{8}\cos(2\omega n) + \frac{1}{8}\cos(4\omega n)$$

$$x^5[n] = \frac{1}{16}\cos(-3\omega n) + \frac{4}{16}\cos(-\omega n) + \frac{6}{16}\cos(\omega n) + \frac{4}{16}\cos(3\omega n) + \frac{1}{16}\cos(5\omega n)$$

and so on. The numerators of the fractions will be recognized as Pascal's triangle. The Central Limit Theorem of probability implies that each kth row can be approximated by a Gaussian curve whose standard deviation (a measure of width) is proportional to the square root of k.

The negative-frequency terms (which have been shown separately here for clarity) are to be combined with the positive ones; the spectral envelope is folded into itself in the same way as in the ring modulation example of Figure 5.4.

As long as the coefficients f_k are all positive numbers or zero, then so are all the amplitudes of the sinusoids in the expansions above. In this case all the phases stay coherent as a varies and so we get a widening of the spectrum (and possibly a drastically increasing amplitude) with increasing values of a. On the other hand, if some of the f_k are positive and others negative, the different expansions will interfere destructively; this will give a more complicated-sounding spectral evolution.

Note also that the successive expansions all contain only even or only odd partials. If the transfer function (in series form) happens to contain only even powers:

$$f(x) = f_0 + f_2 x^2 + f_4 x^4 + \cdots$$

then the result, having only even partials, will sound an octave higher than the incoming sinusoid. If only odd powers show up in the expansion of $f(x)$, then the output will contain only odd partials. Even if f can't be expressed exactly as a power series (for example, the clipping function of Figure 5.3), it is still true that if f is an even function, i.e., if

$$f(-x) = f(x)$$

you will get only even harmonics and if f is an odd function,

$$f(-x) = -f(x)$$

you will get odd harmonics.

Many mathematical tricks have been proposed to use waveshaping to generate specified spectra. It turns out that you can generate pure sinusoids at any harmonic of the fundamental by using a Chebychev polynomial as a transfer function [Leb79] [DJ85], and from there you can go on to build any desired static spectrum (Example E05.chebychev.pd demonstrates this.) Generating *families* of spectra by waveshaping a sinusoid of variable amplitude turns out to be trickier, although several interesting special cases have been found, some of which are developed in detail in Chapter 6.

5.4   Frequency and Phase Modulation

If a sinusoid is given a frequency which varies slowly in time we hear it as having a varying pitch. But if the pitch changes so quickly that our ears can't track the change—for instance, if the change itself occurs at or above the fundamental frequency of the sinusoid—we hear a timbral change. The timbres so generated are rich and widely varying. The discovery by John Chowning of this possibility [Cho73] revolutionized the field of computer music. Here we develop *frequency modulation*, usually called *FM*, as a

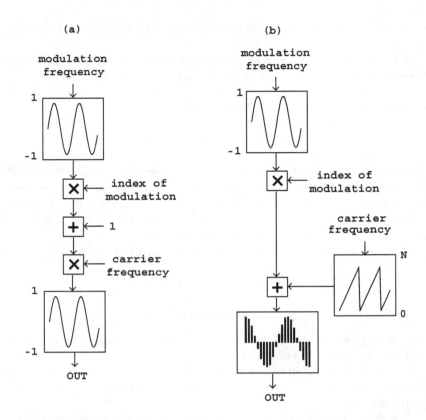

Figure 5.8: Block diagram for frequency modulation (FM) synthesis: (a) the classic form; (b) realized as phase modulation.

special case of waveshaping [Leb79] [DJ85, pp.155-158]; the analysis given here is somewhat different [Puc01].

The FM technique, in its simplest form, is shown in Figure 5.8 (part a). A frequency-modulated sinusoid is one whose frequency varies sinusoidally, at some angular frequency ω_m, about a central frequency ω_c, so that the instantaneous frequencies vary between $(1-r)\omega_c$ and $(1+r)\omega_c$, with parameters ω_m controlling the frequency of variation, and r controlling the depth of variation. The parameters ω_c, ω_m, and r are called the *carrier frequency*, the *modulation frequency*, and the *index of modulation*, respectively.

It is customary to use a simpler, essentially equivalent formulation in which the phase, instead of the frequency, of the carrier sinusoid is modulated sinusoidally. (This gives an equivalent result since the instantaneous frequency is the rate of change of phase, and since the rate of change of a sinusoid is just another sinusoid.) The phase modulation formulation is shown in part (b) of the figure.

We can analyze the result of phase modulation as follows, assuming that the modulating oscillator and the wavetable are both sinusoidal, and that the carrier and modulation frequencies don't themselves vary in time. The resulting signal can then be written as

$$x[n] = \cos(a\cos(\omega_m n) + \omega_c n)$$

The parameter a, which takes the place of the earlier parameter r, is likewise called the index of modulation; it too controls the extent of frequency variation relative to the carrier frequency ω_c. If $a = 0$, there is no frequency variation and the expression reduces to the unmodified, carrier sinusoid; as a increases the waveform becomes more complex.

To analyse the resulting spectrum we can rewrite the signal as,

$$x[n] = \cos(\omega_c n) * \cos(a\cos(\omega_m n))$$

$$- \sin(\omega_c n) * \sin(a\cos(\omega_m n))$$

We can consider the result as a sum of two waveshaping generators, each operating on a sinusoid of frequency ω_m and with a waveshaping index a, and each ring modulated with a sinusoid of frequency ω_c. The waveshaping function f is given by $f(x) = \cos(x)$ for the first term and by $f(x) = \sin(x)$ for the second.

Returning to Figure 5.4, we can predict what the spectrum will look like. The two harmonic spectra, of the waveshaping outputs

$$\cos(a\cos(\omega_m n))$$

and

$$\sin(a\cos(\omega_m n))$$

have, respectively, harmonics tuned to

$$0, 2\omega_m, 4\omega_m, \ldots$$

and

$$\omega_m, 3\omega_m, 5\omega_m, \ldots$$

and each is multiplied by a sinusoid at the carrier frequency. So there will be a spectrum centered at the carrier frequency ω_c, with sidebands at both even and odd multiples of the modulation frequency ω_m, contributed respectively by the sine and cosine waveshaping terms above. The index of modulation a, as it changes, controls the relative strength of the various partials. The partials themselves are situated at the frequencies

$$\omega_c + m\omega_m$$

where

$$m = \ldots - 2, -1, 0, 1, 2, \ldots$$

As with any situation where two periodic signals are multiplied, if there is some common supermultiple of the two periods, the resulting product will repeat at that longer period. So if the two periods are $k\tau$ and $m\tau$, where k and m are relatively prime, they both repeat after a time interval of $km\tau$. In other words, if the two have frequencies which are both multiples of some common frequency, so that $\omega_m = k\omega$ and $\omega_c = m\omega$, again with k and m relatively prime, the result will repeat at a frequency of the common submultiple ω. On the other hand, if no common submultiple ω can be found, or if the only submultiples are lower than any discernible pitch, then the result will be inharmonic.

Much more about FM can be found in textbooks [Moo90, p. 316] [DJ85, pp.115-139] [Bou00] and the research literature. Some of the possibilities are shown in the following examples.

5.5 Examples

Ring modulation and spectra

Example E01.spectrum.pd serves to introduce a spectrum measurement tool we'll be using; here we'll skip to the second example, E02.ring.modulation.pd, which shows the effect of ring modulating a harmonic spectrum (which was worked out theoretically in Section 5.2 and shown in Figure 5.4). In the example we consider a signal whose harmonics (from 0 through 5) all have unit amplitude. The harmonics may be turned on and off separately using toggle switches. When they are all on, the spectral envelope peaks at DC (because the constant signal counts twice as strongly as the other sinusoids), has a flat region from harmonics 1 through 5, and then descends to zero.

In the signal generation portion of the patch (part (a) of the figure), we sum the six partials and multiply the sum by the single, carrier oscillator. (The six signals are summed implicitly by connecting them all to the same inlet of the *~ object.) The value of "fundamental" at the top is computed to line up well with the spectral analysis, whose result is shown in part (b) of the figure.

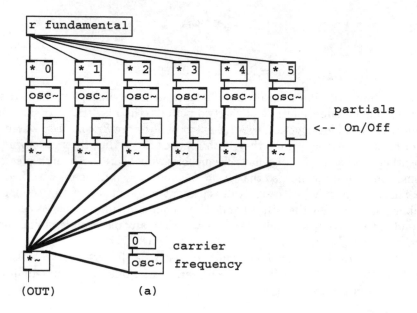

(OUT)   (a)

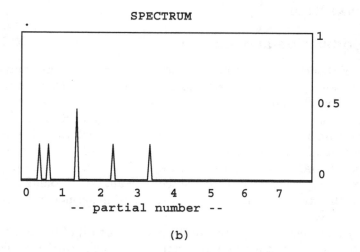

Figure 5.9: Ring modulation of a complex tone by a sinusoid: (a) its realization; (b) a measured spectrum.

The spectral analysis (which uses techniques that won't be described until Chapter 9) shows the location of the sinusoids (assuming a discrete spectrum) on the horizontal axis and their magnitudes on the vertical one. So the presence of a peak at DC of magnitude one in the spectrum of the input signal predicts, à la Figure 5.3, that there should be a peak in the output spectrum, at the carrier frequency, of height $1/2$. Similarly, the two other sinusoids in the input signal, which have height $1/2$ in the spectrum, give rise to two peaks each, of height $1/4$, in the output. One of these four has been reflected about the left edge of the figure (taking the absolute value of its negative frequency).

Octave divider and formant adder

As suggested in Section 5.2, when considering the result of modulating a complex harmonic (i.e., periodic) signal by a sinusoid, an interesting special case is to set the carrier oscillator to $1/2$ the fundamental frequency, which drops the resulting sound an octave with only a relatively small deformation of the spectral envelope. Another is to modulate by a sinusoid at several times the fundamental frequency, which in effect displaces the spectral envelope without changing the fundamental frequency of the result. This is demonstrated in Example E03.octave.divider.pd (Figure 5.10). The signal we process here is a recorded, spoken voice.

The subpatches **pd looper** and **pd delay** hide details. The first is a looping sampler as introduced in Chapter 2. The second is a delay of 1024 samples, which uses objects that are introduced later in Chapter 7. We will introduce one object class here:

| **fiddle~** |: pitch tracker. The inlet takes a signal to analyze, and messages to change settings. Depending on its creation arguments **fiddle~** may have a variable number of outlets offering various information about the input signal. As shown here, with only one creation argument to specify window size, the third outlet attempts to report the pitch of the input, and the amplitude of that portion of the input which repeats (at least approximately) at the reported pitch. These are output as a list of two numbers. The pitch, which is in MIDI units, is reported as zero if none could be identified.

In this patch the third outlet is unpacked into its pitch and amplitude components, and the pitch component is filtered by the **moses** object so that only successful pitch estimates (nonzero ones) are used. These are converted to units of frequency by the **mtof** object. Finally, the frequency estimates are either reduced by $1/2$ or else multiplied by 15, depending on the selected multiplier, to provide the modulation frequency. In the first case we get an octave divider, and in the second, additional high harmonics that deform the vowels.

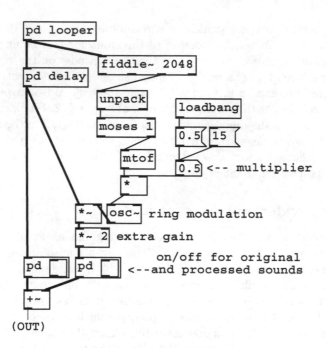

Figure 5.10: Lowering the pitch of a sound by an octave by determining its pitch and modulating at half the fundamental.

Waveshaping and difference tones

Example E04.difference.tone.pd (Figure 5.11) introduces waveshaping, demonstrating the nonlinearity of the process. Two sinusoids (300 and 225 Hertz, or a ratio of 4 to 3) are summed and then clipped, using a new object class:

`clip~` : signal clipper. When the signal lies between the limits specified by the arguments to the `clip~` object, it is passed through unchanged; but when it falls below the lower limit or rises above the upper limit, it is replaced by the limit. The effect of clipping a sinusoidal signal was shown graphically in Figure 5.6.

As long as the amplitude of the sum of sinusoids is less than 50 percent, the sum can't exceed one in absolute value and the `clip~` object passes the pair of sinusoids through unchanged to the output. As soon as the amplitude exceeds 50 percent, however, the nonlinearity of the `clip~` object brings forth distortion products (at frequencies $300m + 225n$ for integers m and n), all of which happening to be multiples of 75, which is thus the

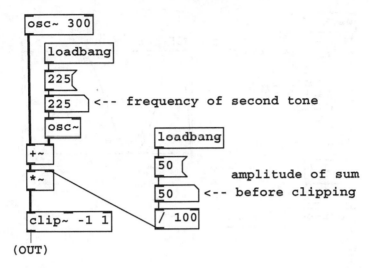

Figure 5.11: Nonlinear distortion of a sum of two sinusoids to create a difference tone.

fundamental of the resulting tone. Seen another way, the shortest common period of the two sinusoids is 1/75 second (which is four periods of the 300 Hertz, tone and three periods of the 225 Hertz tone), so the result repeats 75 times per second.

The frequency of the 225 Hertz tone in the patch may be varied. If it is moved slightly away from 225, a beating sound results. Other values find other common subharmonics, and still others give rise to rich, inharmonic tones.

Waveshaping using Chebychev polynomials

Example E05.chebychev.pd (Figure 5.12) demonstrates how you can use waveshaping to generate pure harmonics. We'll limit ourselves to a specific example here in which we would like to generate the pure fifth harmonic,

$$\cos(5\omega n)$$

by waveshaping a sinusoid

$$x[n] = \cos(\omega n)$$

We need to find a suitable transfer function $f(x)$. First we recall the formula for the waveshaping function $f(x) = x^5$ (Page 131), which gives first, third

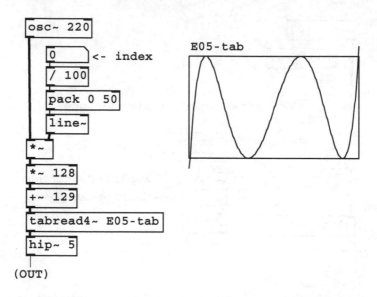

Figure 5.12: Using Chebychev polynomials as waveshaping transfer functions.

and fifth harmonics:

$$16x^5 = \cos(5\omega n) + 5\cos(3\omega n) + 10\cos(\omega n)$$

Next we add a suitable multiple of x^3 to cancel the third harmonic:

$$16x^5 - 20x^3 = \cos(5\omega n) - 5\cos(\omega n)$$

and then a multiple of x to cancel the first harmonic:

$$16x^5 - 20x^3 + 5x = \cos(5\omega n)$$

So for our waveshaping function we choose

$$f(x) = 16x^5 - 20x^3 + 5x$$

This procedure allows us to isolate any desired harmonic; the resulting functions f are known as *Chebychev polynomials* [Leb79].

To incorporate this in a waveshaping instrument, we simply build a patch that works as in Figure 5.5, computing the expression

$$x[n] = f(a[n]\cos(\omega n))$$

where $a[n]$ is a suitable *index* which may vary as a function of the sample number n. When a happens to be one in value, out comes the pure fifth harmonic. Other values of a give varying spectra which, in general, have first and third harmonics as well as the fifth.

By suitably combining Chebychev polynomials we can fix any desired superposition of components in the output waveform (again, as long as the waveshaping index is one). But the real promise of waveshaping—that by simply changing the index we can manufacture spectra that evolve in interesting but controllable ways—is not addressed, at least directly, in the Chebychev picture.

Waveshaping using an exponential function

We return again to the spectra computed on Page 131, corresponding to waveshaping functions of the form $f(x) = x^k$. We note with pleasure that not only are they all in phase (so that they can be superposed with easily predictable results) but also that the spectra spread out as k increases. Also, in a series of the form,

$$f(x) = f_0 + f_1 x + f_2 x^2 + \cdots$$

a higher index of modulation will lend more relative weight to the higher power terms in the expansion; as we saw seen earlier, if the index of modulation is a, the various x^k terms are multiplied by f_0, af_1, $a^2 f_2$, and so on.

Now suppose we wish to arrange for different terms in the above expansion to dominate the result in a predictable way as a function of the index a. To choose the simplest possible example, suppose we wish f_0 to be the largest term for $0 < a < 1$, then for it to be overtaken by the more quickly growing af_1 term for $1 < a < 2$, which is then overtaken by the $a^2 f_2$ term for $2 < a < 3$ and so on, so that each nth term takes over at index n. To make this happen we just require that

$$f_1 = f_0, 2f_2 = f_1, 3f_3 = f_2, \ldots$$

and so choosing $f_0 = 0$, we get $f_1 = 1$, $f_2 = 1/2$, $f_3 = 1/6$, and in general,

$$f_k = \frac{1}{1 \cdot 2 \cdot 3 \cdot \ldots \cdot k}$$

These happen to be the coefficients of the power series for the function

$$f(x) = e^x$$

where $e \approx 2.7$ is Euler's constant.

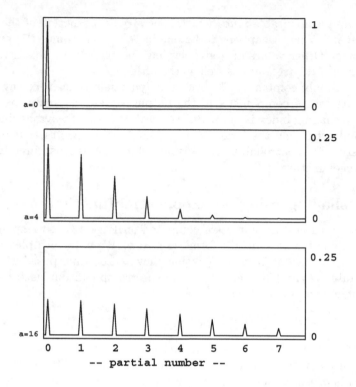

Figure 5.13: Spectra of waveshaping output using an exponential transfer function. Indices of modulation of 0, 4, and 16 are shown; note the different vertical scales.

Before plugging in e^x as a transfer function it's wise to plan how we will deal with signal amplitude, since e^x grows quickly as x increases. If we're going to plug in a sinusoid of amplitude a, the maximum output will be e^a, occurring whenever the phase is zero. A simple and natural choice is simply to divide by e^a to reduce the peak to one, giving:

$$f(a\cos(\omega n)) = \frac{e^{a\cos(\omega n)}}{e^a} = e^{a(\cos(\omega n)-1)}$$

This is realized in Example E06.exponential.pd. Resulting spectra for $a = 0$, 4, and 16 are shown in Figure 5.13. As the waveshaping index rises, progressively less energy is present in the fundamental; the energy is increasingly spread over the partials.

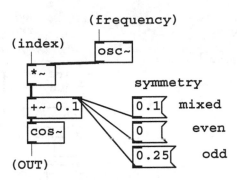

Figure 5.14: Using an additive offset to a cosine transfer function to alter the symmetry between even and odd. With no offset the symmetry is even. For odd symmetry, a quarter cycle is added to the phase. Smaller offsets give a mixture of even and odd.

Sinusoidal waveshaping: evenness and oddness

Another interesting class of waveshaping transfer functions is the sinusoids:

$$f(x) = \cos(x + \phi)$$

which include the cosine and sine functions (got by choosing $\phi = 0$ and $\phi = -\pi/2$, respectively). These functions, one being even and the other odd, give rise to even and odd harmonic spectra, which turn out to be:

$$\cos(a\cos(\omega n)) = J_0(a) - 2J_2(a)\cos(2\omega n) + 2J_4(a)\cos(4\omega n) - 2J_6(a)\cos(6\omega n) \pm \cdots$$

$$\sin(a\cos(\omega n)) = 2J_1(a)\cos(\omega n) - 2J_3(a)\cos(3\omega n) + 2J_5(a)\cos(5\omega n) \mp \cdots$$

The functions $J_k(a)$ are the *Bessel functions* of the first kind, which engineers sometimes use to solve problems about vibrations or heat flow on discs. For other values of ϕ, we can expand the expression for f:

$$f(x) = \cos(x)\cos(\phi) - \sin(x)\sin(\phi)$$

so the result is a mix between the even and the odd harmonics, with ϕ controlling the relative amplitudes of the two. This is demonstrated in Patch E07.evenodd.pd, shown in Figure 5.14.

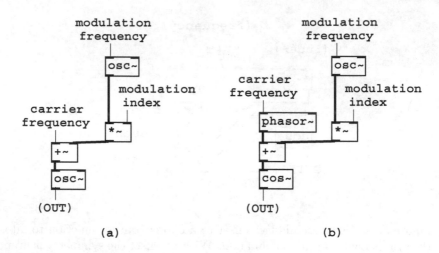

Figure 5.15: Pd patches for: (a) frequency modulation; (b) phase modulation.

Phase modulation and FM

Example E08.phase.mod.pd, shown in Figure 5.15, shows how to use Pd to realize true frequency modulation (part a) and phase modulation (part b). These correspond to the block diagrams of Figure 5.8. To accomplish phase modulation, the carrier oscillator is split into its phase and cosine lookup components. The signal is of the form

$$x[t] = \cos(\omega_c n + a \cos(\omega_m n))$$

where ω_c is the carrier frequency, ω_m is the modulation frequency, and a is the index of modulation—all in angular units.

We can predict the spectrum by expanding the outer cosine:

$$x[t] = \cos(\omega_c n) \cos(a \cos(\omega_m n)) - \sin(\omega_c n) \sin(a \cos(\omega_m n))$$

Plugging in the expansions from Page 143 and simplifying yields:

$$x[t] = J_0(a) \cos(\omega_c n)$$

$$+ J_1(a) \cos((\omega_c + \omega_m)n + \frac{\pi}{2}) + J_1(a) \cos((\omega_c - \omega_m)n + \frac{\pi}{2})$$

$$+ J_2(a) \cos((\omega_c + 2\omega_m)n + \pi) + J_2(a) \cos((\omega_c - 2\omega_m)n + \pi)$$

$$+ J_3(a) \cos((\omega_c + 3\omega_m)n + \frac{3\pi}{2}) + J_3(a) \cos((\omega_c - 3\omega_m)n + \frac{3\pi}{2}) + \cdots$$

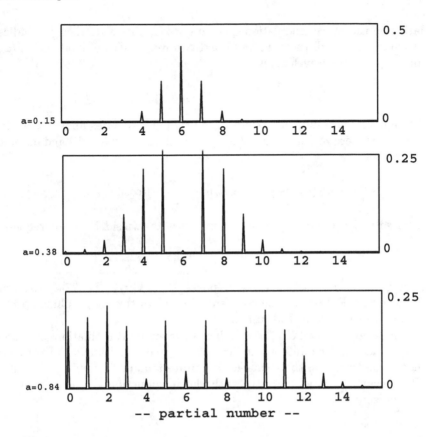

Figure 5.16: Spectra from phase modulation at three different indices. The indices are given as multiples of 2π radians.

So the components are centered about the carrier frequency ω_c with sidebands extending in either direction, each spaced ω_m from the next. The amplitudes are functions of the index of modulation, and don't depend on the frequencies. Figure 5.16 shows some two-operator phase modulation spectra, measured using Example E09.FM.spectrum.pd.

Phase modulation can thus be seen simply as a form of ring modulated waveshaping. So we can use the strategies described in Section 5.2 to generate particular combinations of frequencies. For example, if the carrier frequency is half the modulation frequency, you get a sound with odd harmonics exactly as in the octave dividing example (Figure 5.10).

Frequency modulation need not be restricted to purely sinusoidal carrier or modulation oscillators. One well-trodden path is to effect phase modulation on the phase modulation spectrum itself. There are then two indices of modulation (call them a and b) and two frequencies of modulation (ω_m and ω_p) and the waveform is:

$$x[n] = \cos(\omega_c n + a\cos(\omega_m n) + b\cos(\omega_p n))$$

To analyze the result, just rewrite the original FM series above, replacing $\omega_c n$ everywhere with $\omega_c n + b\cos(\omega_p n)$. The third positive sideband becomes for instance:

$$J_3(a)\cos((\omega_c + 3\omega_m)n + \frac{3\pi}{2} + b\cos(\omega_p n))$$

This is itself just another FM spectrum, with its own sidebands of frequency

$$\omega_c + 3\omega_m + k\omega_p, k = 0, \pm1, \pm2, \ldots$$

having amplitude $J_3(a)J_k(b)$ and phase $(3 + k)\pi/2$ [Leb77]. Example E10.complex.FM.pd (not shown here) illustrates this by graphing spectra from a two-modulator FM instrument.

Since early times [Sch77] researchers have sought combinations of phases, frequencies, and modulation indices, for simple and compact phase modulation instruments, that manage to imitate familiar instrumental sounds. This became a major industry with the introduction of commercial FM synthesizers.

Exercises

1. A sound has fundamental 440. How could it be ring modulated to give a tone at 110 Hertz with only odd partials? How could you then fill in the even ones if you wanted to?

2. A sinusoid with frequency 400 and unit peak amplitude is squared. What are the amplitudes and frequencies of the new signal's components?

3. What carrier and modulation frequencies would you give a two-operator FM instrument to give frequencies of 618, 1000, and 2618 Hertz? (This is a prominent feature of Chowning's *Stria* [DJ85].)

4. Two sinusoids with frequency 300 and 400 Hertz and peak amplitude one (so RMS amplitude ≈ 0.707) are multiplied. What is the RMS amplitude of the product?

5. Suppose you wanted to make FM yet more complicated by modulating the *modulating* oscillator, as in:

$$\cos(\omega_c n + a \cos(\omega_m n + b \cos(\omega_p n)))$$

How, qualitatively speaking, would the spectrum differ from that of the simple two-modulator example (Section 5.5)?

0. A sinusoid at a frequency ω is ring modulated by another sinusoid at exactly the same frequency. At what phase differences will the DC component of the result disappear?

Chapter 6

Designer Spectra

As suggested at the beginning of the previous chapter, a powerful way to synthesize musical sounds is to specify—and then realize—specific trajectories of pitch (or more generally, frequencies of partials), along with trajectories of spectral envelope [Puc01]. The spectral envelope is used to determine the amplitude of the individual partials, as a function of their frequencies, and is thought of as controlling the sound's (possibly time-varying) timbre.

A simple example of this would be to imitate a plucked string by constructing a sound with harmonically spaced partials in which the spectral envelope starts out rich but then dies away exponentially with higher frequencies decaying faster than lower ones, so that the timbre mellows over time. Spectral-evolution models for various acoustic instruments have been proposed [GM77] [RM69] . A more complicated example is the spoken or sung voice, in which vowels appear as spectral envelopes, dipthongs and many consonants appear as time variations in the spectral envelopes, and other consonants appear as spectrally shaped noise.

Spectral envelopes may be obtained from analysis of recorded sounds (developed in Chapter 9) or from purely synthetic criteria. To specify a spectral envelope from scratch for every possible frequency would be tedious, and in most cases you would want to describe them in terms of their salient features. The most popular way of doing this is to specify the size and shape of the spectral envelope's peaks, which are called *formants*. Figure 6.1 shows a spectral envelope with two formants. Although the shapes of the two peaks in the spectral envelope are different, they can both be roughly described by giving the coordinates of each apex (which give the formant's *center frequency* and amplitude) and each formant's *bandwidth*. A typical measure of bandwidth would be the width of the peak at a level 3 decibels below its apex. Note that if the peak is at (or near) the $f = 0$

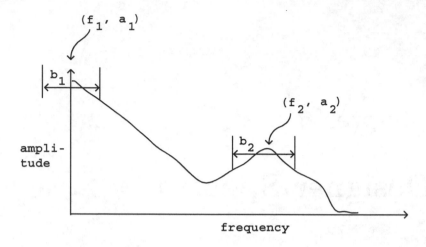

Figure 6.1: A spectral envelope showing the frequencies, amplitudes, and bandwidths of two formants.

axis, we pretend it falls off to the left at the same rate as (in reality) it falls off to the right.

Suppose we wish to generate a harmonic sound with a specified collection of formants. Independently of the fundamental frequency desired, we wish the spectrum to have peaks with prescribed center frequencies, amplitudes, and bandwidths. Returning to the phase modulation spectra shown in Figure 5.16, we see that, at small indices of modulation at least, the result has a single, well-defined spectral peak. We can imagine adding several of these, all sharing a fundamental (modulating) frequency but with carriers tuned to different harmonics to select the various desired center frequencies, and with indices of modulation chosen to give the desired bandwidths. This was first explored by Chowning [Cho89] who arranged formants generated by phase modulation to synthesize singing voices. In this chapter we'll establish a general framework for building harmonic spectra with desired, possibly time-varying, formants.

6.1   Carrier/Modulator Model

Earlier we saw how to use ring modulation to modify the spectrum of a periodic signal, placing spectral peaks in specified locations (see Figure 5.4, Page 126). To do so we need to be able to generate periodic signals whose spectra have maxima at DC and fall off monotonically with

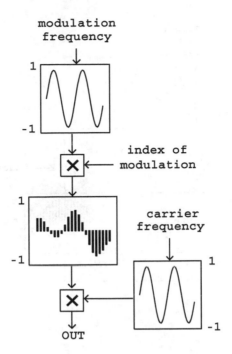

Figure 6.2: Ring modulated waveshaping for formant generation.

increasing frequency. If we can make a signal with a formant at frequency zero—and no other formants besides that one—we can use ring modulation to displace the formant to any desired harmonic. If we use waveshaping to generate the initial formant, the ring modulation product will be of the form

$$x[n] = \cos(\omega_c n) f(a \cos(\omega_m n))$$

where ω_c (the *carrier frequency*) is set to the formant center frequency and $f(a \cdot \cos(\omega_m n))$ is a signal with fundamental frequency determined by ω_m, produced using a waveshaping function f and index a. This second term is the signal we wish to give a formant at DC with a controllable bandwidth. A block diagram for synthesizing this signal is shown in Figure 6.2.

Much earlier in Section 2.4 we introduced the technique of *timbre stretching*, as part of the discussion of wavetable synthesis. This technique, which is capable of generating complex, variable timbres, can be fit into the same framework. The enveloped wavetable output for one cycle is:

$$x(\phi) = T(c\phi) * W(a\phi)$$

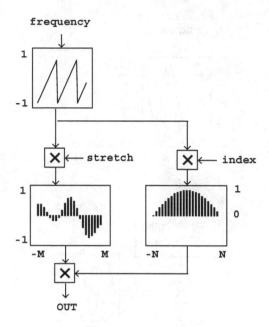

Figure 6.3:   Wavetable synthesis generalized as a variable spectrum generator.

where ϕ, the phase, satisfies $-\pi \leq \phi \leq \pi$. Here T is a function stored in a wavetable, W is a windowing function, and c and a are the wavetable stretching and a modulation index for the windowing function. Figure 6.3 shows how to realize this in block diagram form. Comparing this to Figure 2.7, we see that the only significant new feature is the addition of the index a.

In this setup, as in the previous one, the first term specifies the placement of energy in the spectrum—in this case, with the parameter c acting to stretch out the wavetable spectrum. This is the role that was previously carried out by the choice of ring modulation carrier frequency ω_c.

Both of these (ring modulated waveshaping and stretched wavetable synthesis) can be considered as particular cases of a more general approach which is to compute functions of the form,

$$x[n] = c(\omega n)m_a(\omega n)$$

where c is a periodic function describing the carrier signal, and m_a is a periodic modulator function which depends on an index a. The modulation functions we're interested in will usually take the form of pulse trains, and the index a will control the width of the pulse; higher values of a will give narrower pulses. In the wavetable case, the modulation function must reach zero at phase wraparound points to suppress any discontinuities in the carrier function when the phase wraps around. The carrier signal will give rise to a single spectral peak (a formant) in the ring modulated waveshaping case; for wavetables, it may have a more complicated spectrum.

In the next section we will further develop the two forms of modulating signal we've introduced here, and in the following one we'll look more closely at the carrier signal.

6.2 Pulse Trains

Pulse trains may be generated either using the waveshaping formulation or the stretched wavetable one. The waveshaping formulation is easier to analyze and control, and we'll consider it first.

6.2.1 Pulse trains via waveshaping

When we use waveshaping the shape of the formant is determined by a modulation term

$$m_a[n] = f(a\cos(\omega n))$$

For small values of the index a, the modulation term varies only slightly from the constant value $f(0)$, so most of the energy is concentrated at DC. As a increases, the energy spreads out among progressively higher harmonics of the fundamental ω. Depending on the function f, this spread may be orderly or disorderly. An orderly spread may be desirable and then again may not, depending on whether our goal is a predictable spectrum or a wide range of different (and perhaps hard-to-predict) spectra.

The waveshaping function $f(x) = e^x$, analyzed on Page 141, gives well-behaved, simple and predictable results. After normalizing suitably, we got the spectra shown in Figure 5.13. A slight rewriting of the waveshaping modulator for this choice of f (and taking the renormalization into account) gives:

$$m_a[n] = e^{a \cdot (\cos(\omega n) - 1))}$$

$$= e^{-\left[b\sin\frac{\omega}{2}\right]^2}$$

where $b^2 = 2a$ so that b is proportional to the bandwidth. This can be rewritten as

$$m_a[n] = g(b\sin\frac{\omega}{2}n)$$

with
$$g(x) = e^{-x^2}$$

Except for a missing normalization factor, this is a Gaussian distribution, sometimes called a "bell curve". The amplitudes of the harmonics are given by Bessel "I" type functions.

Another fine choice is the (again unnormalized) Cauchy distribution:

$$h(x) = \frac{1}{1 + x^2}$$

which gives rise to a spectrum of exponentially falling harmonics:

$$h(b\sin(\frac{\omega}{2}n)) = G \cdot \left(\frac{1}{2} + H\cos(\omega n) + H^2\cos(2\omega n) + \cdots \right)$$

where G and H are functions of the index b (explicit formulas are given in [Puc95a]).

In both this and the Gaussian case above, the bandwidth (counted in peaks, i.e., units of ω) is roughly proportional to the index b, and the amplitude of the DC term (the apex of the spectrum) is roughly proportional to $1/(1+b)$. For either waveshaping function (g or h), if b is larger than about 2, the waveshape of $m_a(\omega n)$ is approximately a (forward or backward) scan of the transfer function, so the resulting waveform looks like pulses whose widths decrease as the specified bandwidth increases.

6.2.2   Pulse trains via wavetable stretching

In the wavetable formulation, a pulse train can be made by a stretched wavetable:

$$M_a(\phi) = W(a\phi)$$

where $-\pi \leq \phi \leq \pi$ is the phase, i.e., the value ωn wrapped to lie between $-\pi$ and π. The function W should be zero at and beyond the points $-\pi$ and π, and rise to a maximum at 0. A possible choice for the function W is

$$W(\phi) = \frac{1}{2}\left(\cos(\phi) + 1\right)$$

which is graphed in part (a) of Figure 6.4. This is known as the *Hann window function*; it will come up again in Chapter 9.

Realizing this as a repeating waveform, we get a succession of (appropriately sampled) copies of the function W, whose duty cycle is $1/a$ (parts b and c of the figure). If you don't wish the copies to overlap the index a must be at least 1. If you want to allow overlap the simplest strategy is to duplicate the block diagram (Figure 6.3) out of phase, as described in Section 2.4 and realized in Section 2.6.

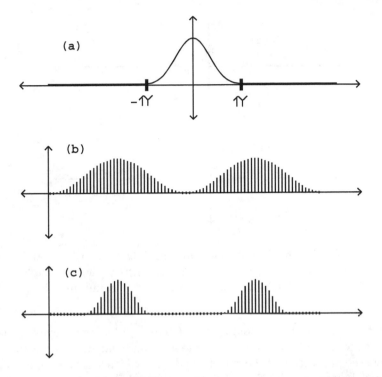

Figure 6.4: Pulse width modulation using the Hann window function: (a) the function $W(\phi) = (1 + \cos(\phi))/2$; (b) the function as a waveform, repeated at a duty cycle of 100% (modulation index $a = 1$); (c) the waveform at a 50% duty cycle ($a = 2$).

6.2.3   Resulting spectra

Before considering more complicated carrier signals to go with the modulators we've seen so far, it is instructive to see what multiplication by a pure sinusoid gives us as waveforms and spectra. Figure 6.5 shows the result of multiplying two different pulse trains by a sinusoid at the sixth partial:

$$\cos(6\omega n)M_a(\omega n)$$

where the index of modulation a is two in both cases. In part (a) M_a is the stretched Hann windowing function; part (b) shows waveshaping via the unnormalized Cauchy distribution. One period of each waveform is shown.

In both situations we see, in effect, the sixth harmonic (the carrier signal) enveloped into a *wave packet* centered at the middle of the cycle,

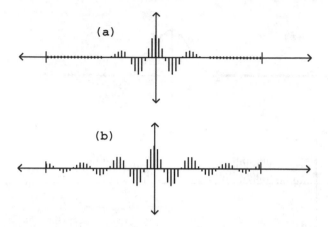

Figure 6.5: Audio signals resulting from multiplying a cosine (partial number 6) by pulse trains: (a) windowing function from the wavetable formulation; (b) waveshaping output using the Cauchy lookup function.

where the phase of the sinusoid is zero. Changing the frequency of the sinusoid changes the center frequency of the formant; changing the width of the packet (the proportion of the waveform during which the sinusoid is strong) changes the bandwidth. Note that the stretched Hann window function is zero at the beginning and end of the period, unlike the waveshaping packet.

Figure 6.6 shows how the shape of the formant depends on the method of production. The stretched wavetable form (part (a) of the figure) behaves well in the neighborhood of the peak, but somewhat oddly starting at four partials' distance from the peak, past which we see what are called *sidelobes*: spurious extra peaks at lower amplitude than the central peak. As the analysis of Section 2.4 predicts, the entire formant, sidelobes and all, stretches or contracts inversely as the pulse train is contracted or stretched in time.

The first, strongest sidelobes on either side are about 32 dB lower in amplitude than the main peak. Further sidelobes drop off slowly when expressed in decibels; the amplitudes decrease as the square of the distance from the center peak so that the sixth sidelobe to the right, three times further than the first one from the center frequency, is about twenty decibels further down. The effect of these sidelobes is often audible as a slight buzziness in the sound.

This formant shape may be made arbitrarily fat (i.e., high bandwidth), but there is a limit on how thin it can be made, since the duty cycle of the waveform cannot exceed 100%. At this maximum duty cycle the formant

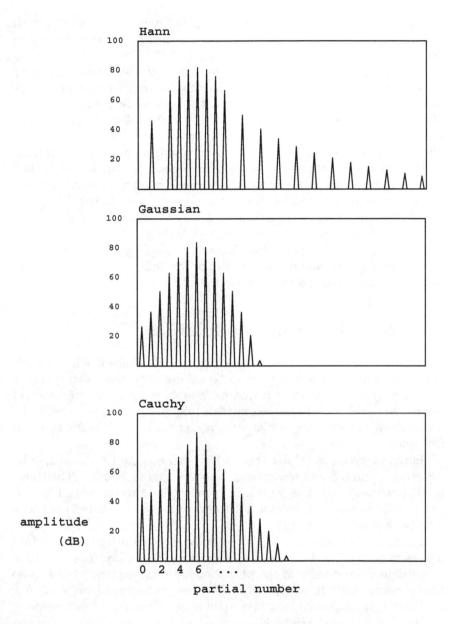

Figure 6.6: Spectra of three ring-modulated pulse trains: (a) the Hann window function, 50% duty cycle (corresponding to an index of 2); (b) a waveshaping pulse train using a Gaussian transfer function; (c) the same, with a Cauchy transfer function. Amplitudes are in decibels.

strength drops to zero at two harmonics' distance from the center peak. If a still lower bandwidth is needed, waveforms may be made to overlap as described in Section 2.6.

Parts (b) and (c) of the figure show formants generated using ring modulated waveshaping, with Gaussian and Cauchy transfer functions. The index of modulation is two in both cases (the same as for the Hann window of part a), and the bandwidth is comparable to that of the Hann example. In these examples there are no sidelobes, and moreover, the index of modulation may be dropped all the way to zero, giving a pure sinusoid; there is no lower limit on bandwidth. On the other hand, since the waveform does not reach zero at the ends of a cycle, this type of pulse train cannot be used to window an arbitrary wavetable, as the Hann pulse train could.

The Cauchy example is particularly handy for designing spectra, since the shape of the formant is a perfect isosceles triangle, when graphed in decibels. On the other hand, the Gaussian example gathers more energy toward the formant, and drops off faster at the tails, and so has a cleaner sound and offers better protection against foldover.

6.3   Movable Ring Modulation

We turn now to the carrier signal, seeking ways to make it more controllable. We would particularly like to be able to slide the spectral energy continuously up and down in frequency. Simply ramping the frequency of the carrier oscillator will not accomplish this, since the spectra won't be harmonic except when the carrier is an integer multiple of the fundamental frequency.

In the stretched wavetable approach we can accomplish this simply by sampling a sinusoid and transposing it to the desired "pitch". The transposed pitch isn't heard as a periodicity since the wavetable itself is read periodically at the fundamental frequency. Instead, the sinusoid is transposed as a spectral envelope.

Figure 6.7 shows a carrier signal produced in this way, tuned to produce a formant centered at 1.5 times the fundamental frequency. The signal has no outright discontinuity at the phase wraparound frequency, but it does have a discontinuity in slope, which, if not removed by applying a suitable modulation signal, would have very audible high-frequency components.

Using this idea we can make a complete description of how to use the block diagram of Figure 6.3 to produce a desired formant. The wavetable lookup on the left hand side would hold a sinusoid (placed symmetrically so that the phase is zero at the center of the wavetable). The right-hand-side wavetable would hold a Hann or other appropriate window function. If we desire the fundamental frequency to be ω, the formant center frequency to

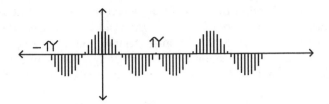

Figure 6.7: Waveform for a wavetable-based carrier signal tuned to 1.5 times the fundamental. Two periods are shown.

be ω_c, and the bandwidth to be ω_b, we set the "stretch" parameter to the *center frequency quotient* defined as ω_c/ω, and the index of modulation to the *bandwidth quotient*, ω_b/ω.

The output signal is simply a sample of a cosine wave at the desired center frequency, repeated at the (unrelated in general) desired period, and windowed to take out the discontinuities at period boundaries.

Although we aren't able to derive this result yet (we will need Fourier analysis), it will turn out that, in the main lobe of the formant, the phases are all zero at the center of the waveform (i.e., the components are all cosines if we consider the phase to be zero at the center of the waveform). This means we may superpose any number of these formants to build a more complex spectrum and the amplitudes of the partials will combine by addition. (The sidelobes don't behave so well: they are alternately of opposite sign and will produce cancellation patterns; but we can often just shrug them off as a small, uncontrollable, residual signal.)

This method leads to an interesting generalization, which is to take a sequence of recorded wavetables, align all their component phases to those of cosines, and use them in place of the cosine function as the carrier signal. The phase alignment is necessary to allow coherent cross-fading between samples so that the spectral envelope can change smoothly. If, for example, we use successive snippets of a vocal sample as input, we get a strikingly effective vocoder; see Section 9.6.

Another technique for making carrier signals that can be slid continuously up and down in frequency while maintaining a fundamental frequency is simply to cross-fade between harmonics. The carrier signal is then:

$$c(\phi) = c(\omega n) = p\cos(k\omega n) + q\cos((k+1)\omega n)$$

where $p + q = 1$ and k is an integer, all three chosen so that

$$(k+q)*\omega = \omega_c$$

so that the spectral center of mass of the two cosines is placed at ω_c. (Note that we make the amplitudes of the two cosines add to one instead of setting the total power to one; we do this because the modulator will operate phase-coherently on them.) To accomplish this we simply set k and q to be the integer and fractional part, respectively, of the center frequency quotient ω_c/ω.

The simplest way of making a control interface for this synthesis technique would be to use ramps to update ω and ω_c, and then to compute q and k as audio signals from the ramped, smoothly varying ω and ω_c. Oddly enough, despite the fact that k, p, and q are discontinuous functions of ω_c/ω, the carrier $c(\phi)$ turns out to vary continuously with ω_c/ω, and so if the desired center frequency ω_c is ramped from value to value the result is a continuous sweep in center frequency. However, more work is needed if discontinuous changes in center frequency are needed. This is not an unreasonable thing to wish for, being analogous to changing the frequency of an oscillator discontinuously.

There turns out to be a good way to accomodate this. The trick to updating k and q is to note that $c(\phi) = 1$ whenever ϕ is a multiple of 2π, regardless of the choice of k, p, and q as long as $p + q = 1$. Hence, we may make discontinuous changes in k, p, and q once per period (right when the phase is a multiple of 2π), without making discontinuities in the carrier signal.

In the specific case of FM, if we wish we can now go back and modify the original formulation to:

$$p \cos(n\omega_2 t + r \cos(\omega_1 t))$$

$$+q \cos((n + 1)\omega_2 t + r \cos(\omega_1 t))$$

This allows us to add glissandi (which are heard as dipthongs) to Chowning's original phase-modulation-based vocal synthesis technique.

6.4   Phase-Aligned Formant (PAF) Generator

Combining the two-cosine carrier signal with the waveshaping pulse generator gives the *phase-aligned formant* generator, usually called by its acronym, PAF. (The PAF is the subject of a 1994 patent owned by IRCAM.) The combined formula is,

$$x[n] = \underbrace{g\left(a \sin(\omega n/2)\right)}_{\text{modulator}} \underbrace{\left[p \cos(k\omega n) + q \cos((k + 1)\omega n)\right]}_{\text{carrier}}$$

Here the function g may be either the Gaussian or Cauchy waveshaping function, ω is the fundamental frequency, a is a modulation index controlling bandwidth, and k, p, and q control the formant center frequency.

Figure 6.8 shows the PAF as a block diagram, separated into a phase generation step, a carrier, and a modulator. The phase generation step outputs a sawtooth signal at the fundamental frequency. The modulator is done by standard waveshaping, with a slight twist added. The formula for the modulator signals in the PAF call for an incoming sinusoid at half the fundamental frequency, i.e., $\sin(\frac{\omega}{2})$, and this nominally would require us to use a phasor tuned to half the fundamental frequency. However, since the waveshaping function is even, we may substitute the absolute value of the sinusoid:

$$\left|\sin(\frac{\omega}{2})\right|$$

which repeats at the frequency ω (the first half cycle is the same as the second one.) We can compute this simply by using a half-cycle sinusoid as a wavetable lookup function (with phase running from $-\pi/2$ to $\pi/2$), and it is this rectified sinusoid that we pass to the waveshaping function.

Although the wavetable function is pictured over both negative and positive values (reaching from -10 to 10), in fact we're only using the positive side for lookup, ranging from 0 to b, the index of modulation. If the index of modulation exceeds the input range of the table (here set to stop at 10 as an example), the table lookup address should be clipped. The table should extend far enough into the tail of the waveshaping function so that the effect of clipping is inaudible.

The carrier signal is a weighted sum of two cosines, whose frequencies are increased by multiplication (by k and $k+1$, respectively) and wrapping. In this way all the lookup phases are controlled by the same sawtooth oscillator.

The quantities k, q, and the wavetable index b are calculated as shown in Figure 6.9. They are functions of the specified fundamental frequency, the formant center frequency, and the bandwidth, which are the original parameters of the algorithm. The quantity p, not shown in the figure, is just $1 - q$.

As described in the previous section, the quantities k, p, and q should only change at phase wraparound points, that is to say, at periods of $2\pi/\omega$. Since the calculation of k, etc., depends on the value of the parameter ω, it follows that ω itself should only be updated when the phase is a multiple of 2π; otherwise, a change in ω could send the center frequency $(k+q)\omega$ to an incorrect value for a (very audible) fraction of a period. In effect, all the parameter calculations should be synchronized to the phase of the original oscillator.

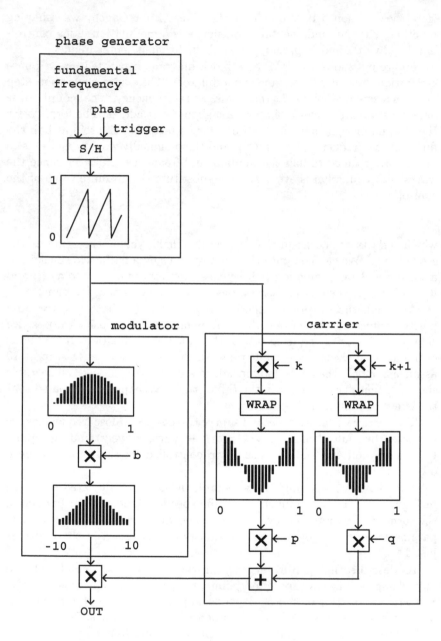

Figure 6.8: The PAF generator as a block diagram.

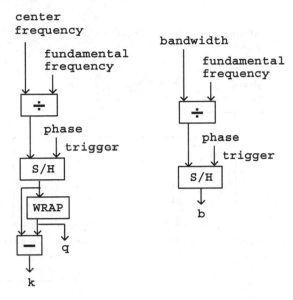

Figure 6.9: Calculation of the time-varying parameters b (the waveshaping index), k, and q for use in the block diagram of Figure 6.8.

Having the oscillator's phase control the updating of its own frequency is an example of *feedback*, which in general means using any process's output as one of its inputs. When processing digital audio signals at a fixed sample rate (as we're doing), it is never possible to have the process's *current* output as an input, since at the time we would need it we haven't yet calculated it. The best we can hope for is to use the previous sample of output—in effect, adding one sample of delay. In block environments (such as Max, Pd, and Csound) the situation becomes more complicated, but we will delay discussing that until the next chapter (and simply wish away the problem in the examples at the end of this one).

The amplitude of the central peak in the spectrum of the PAF generator is roughly $1/(1+b)$; in other words, close to unity when the index b is smaller than one, and falling off inversely with larger values of b. For values of b less than about ten, the loudness of the output does not vary greatly, since the introduction of other partials, even at lower amplitudes, offsets the decrease of the center partial's amplitude. However, if using the PAF to generate formants with specified peak amplitudes, the output should be multiplied by $1 + b$ (or even, if necessary, a better approximation of the correction factor, whose exact value depends on the waveshaping function). This amplitude correction should be ramped, not sampled-and-held.

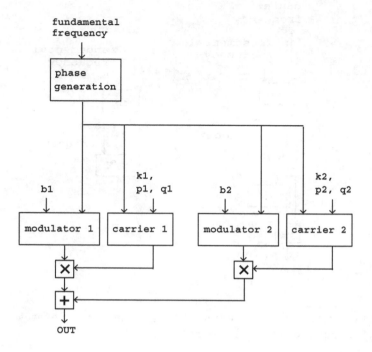

Figure 6.10: Block diagram for making a spectrum with two formants using the PAF generator.

Since the expansion of the waveshaping (modulator) signal consists of all cosine terms (i.e., since they all have initial phase zero), as do the two components of the carrier, it follows from the cosine product formula that the components of the result are all cosines as well. This means that any number of PAF generators, if they are made to share the same oscillator for phase generation, will all be in phase and combining them gives the sum of the individual spectra. So we can make a multiple-formant version as shown in Figure 6.10.

Figure 6.11 shows a possible output of a pair of formants generated this way; the first formant is centered halfway between partials 3 and 4, and the second at partial 12, with lower amplitude and bandwidth. The Cauchy waveshaping function was used, which makes linearly sloped spectra (viewed in dB). The two superpose additively, so that the spectral envelope curves smoothly from one formant to the other. The lower formant also adds to its own reflection about the vertical axis, so that it appears slightly curved upward there.

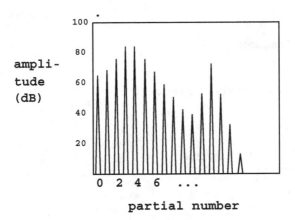

Figure 6.11: Spectrum from a two-formant PAF generator.

The PAF generator can be altered if desired to make inharmonic spectra by sliding the partials upward or downward in frequency. To do this, add a second oscillator to the phase of both carrier cosines, but not to the phase of the modulation portion of the diagram, nor to the controlling phase of the sample-and-hold units. It turns out that the sample-and-hold strategy for smooth parameter updates still works; and furthermore, multiple PAF generators sharing the same phase generation portion will still be in phase with each other.

This technique for superposing spectra does not work as predictably for phase modulation as it does for the PAF generator; the partials of the phase modulation output have complicated phase relationships and they seem difficult to combine coherently. In general, phase modulation will give more complicated patterns of spectral evolution, whereas the PAF is easier to predict and turn to specific desired effects.

6.5 Examples

Wavetable pulse train

Example F01.pulse.pd (Figure 6.12) generates a variable-width pulse train using stretched wavetable lookup. Figure 6.13 shows two intermediate products of the patch and its output. The patch carries out the job in the simplest possible way, placing the pulse at phase π instead of phase zero; in later examples this will be fixed by adding 0.5 to the phase and wrapping.

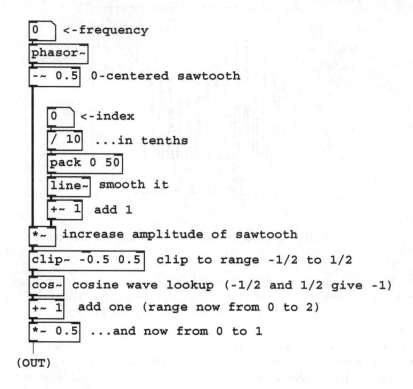

Figure 6.12: Example patch F01.pulse.pd, which synthesizes a pulse train using stretched wavetable lookup.

The initial phase is adjusted to run from -0.5 to 0.5 and then scaled by a multiplier of at least one, resulting in the signal of Figure 6.13 (part a); this corresponds to the output of the *~ object, fifth from bottom in the patch shown. The graph in part (b) shows the result of clipping the sawtooth wave back to the interval between −0.5 and 0.5, using the clip~ object. If the scaling multiplier were at its minimum (one), the sawtooth would only range from -0.5 to 0.5 anyway and the clipping would have no effect. For any value of the scaling multiplier greater than one, the clipping output sits at the value -0.5, then ramps to 0.5, then sits at 0.5. The higher the multiplier, the faster the waveform ramps and the more time it spends clipped at the bottom and top.

The cos~ object then converts this waveform into a pulse. Inputs of both -0.5 and 0.5 go to -1 (they are one cycle apart); at the midpoint of the

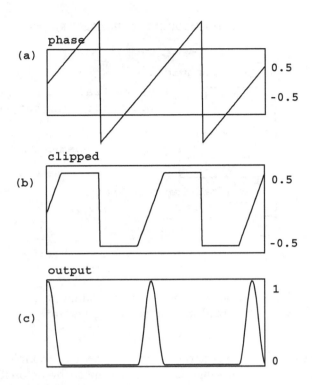

Figure 6.13: Intermediate audio signals from Figure 6.12: (a) the result of multiplying the phasor by the "index"; (b) the same, clipped to lie between -0.5 and 0.5; (c) the output.

waveform, the input is 0 and the output is thus 1. The output therefore sits at -1, traces a full cycle of the cosine function, then comes back to rest at -1. The proportion of time the waveform spends tracing the cosine function is one divided by the multiplier; so it's 100% for a multiplier of 1, 50% for 2, and so on. Finally, the pulse output is adjusted to range from 0 to 1 in value; this is graphed in part (c) of the figure.

Simple formant generator

The next three examples demonstrate the sound of the varying pulse width, graph its spectrum, and contrast the waveshaping pulse generator. Skipping to Example F05.ring.modulation.pd (Figure 6.14), we show the simplest way of combining the pulse generator with a ring modulating oscillator to

RING MODULATED PULSE TRAINS

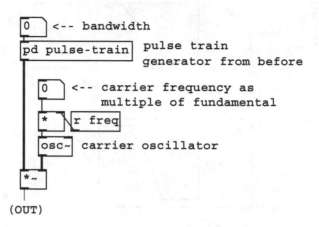

Figure 6.14: Excerpt from example F05.ring.modulation.pd combining ring modulation with a stretched wavetable pulse generator

make a formant. The pulse train from the previous example is contained in the **pd pulse-train** subpatch. It is multiplied by an oscillator whose frequency is controlled as a multiple of the fundamental frequency. If the multiple is an integer, a harmonic sound results. No attempt is made to control the relative phases of the components of the pulse train and of the carrier sinusoid.

The next example, F06.packets.pd (Figure 6.15), shows how to combine the stretched wavetable pulse train with a sampled sinusoid to realize movable formants, as described in Section 6.3. The pulse generator is as before, but now the carrier signal is a broken sinusoid. Since its phase is the fundamental phase times the center frequency quotient, the sample-to-sample phase increment is the same as for a sinusoid at the center frequency. However, when the phase wraps around, the carrier phase jumps to a different place in the cycle, as was illustrated in Figure 6.7. Although the bandwidth quotient ω_b/ω must be at least one, the center frequency quotient ω_c/ω may be as low as zero if desired.

Two-cosine carrier signal

Example F08.two.cosines.pd (Figure 6.16) shows how to make a carrier signal that cross-fades between harmonics to make continuously variable

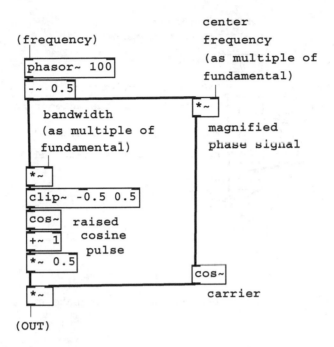

Figure 6.15: Using stretched wavetable synthesis to make a formant with movable center frequency.

center frequencies. The center frequency quotient appears as the output of a line~ object. This is separated into its fractional part (using the wrap~ object) and its integer part (by subtracting the fractional part from the original). These are labeled as q and k to agree with the treatment in Section 6.3.

The phase—a sawtooth wave at the fundamental frequency—is multiplied by both k and $k + 1$ (the latter by adding the original sawtooth into the former), and the cosines of both are taken; they are therefore at k and $k + 1$ times the fundamental frequency and have no discontinuities at phase wrapping points. The next several objects in the patch compute the weighted sum $pc_1 + qc_2$, where c_1, c_2 are the two sinusoids and $p = 1 - q$, by evaluating an equivalent expression, $c_1 + q(c_2 - c_1)$. This gives us the desired movable-frequency carrier signal.

Example F09.declickit.pd (not shown here) shows how, by adding a samphold~ object after the line~ object controlling center frequency, you can avoid discontinuities in the output signal even if the desired center

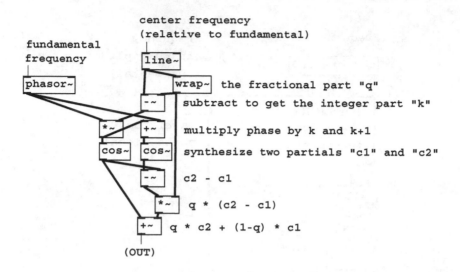

Figure 6.16: Cross-fading between sinusoids to make movable center frequencies.

frequency changes discontinuously. In the example the center frequency quotient alternates between 4 and 13.5. At ramp times below about 20 msec there are audible artifacts when using the line~ object alone which disappear when the samphold~ object is added. (A disadvantage of sample-and-holding the frequency quotient is that, for very low fundamental frequencies, the changes can be heard as discrete steps. So in situations where the fundamental frequency is low and the center frequency need not change very quickly, it may be better to omit the sample-and-hold step.)

The next two examples demonstrate using the crossfading-oscillators carrier as part of the classic two-operator phase modulation technique. The same modulating oscillator is added separately to the phases of the two cosines. The resulting spectra can be made to travel up and down in frequency, but because of the complicated phase relationships between neighboring peaks in the phase modulation spectrum, no matter how you align two such spectra you can never avoid getting phase cancellations where they overlap.

The PAF generator

Example F12.paf.pd (Figure 6.17) is a realization of the PAF generator, described in Section 6.4. The control inputs specify the fundamental fre-

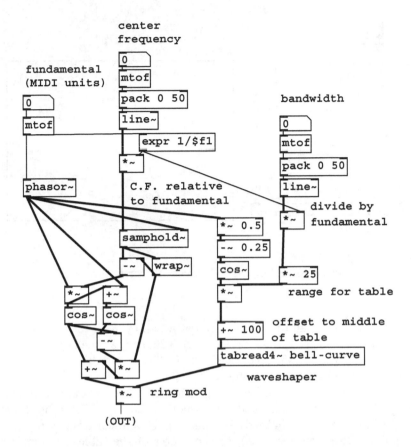

Figure 6.17: The phase-aligned formant (PAF) synthesis algorithm.

quency, the center frequency, and the bandwidth, all in "MIDI" units. The first steps taken in the realization are to divide center frequency by fundamental (to get the center frequency quotient) and bandwidth by fundamental to get the index of modulation for the waveshaper. The center frequency quotient is sampled-and-held so that it is only updated at periods of the fundamental.

The one oscillator (the **phasor~** object) runs at the fundamental frequency. This is used both to control a **samphold~** object which synchronizes updates to the center frequency quotient (labeled "C.F. relative to fundamental" in the figure), and to compute phases for both **cos~** objects which operate as shown earlier in Figure 6.16.

The waveshaping portion of the patch uses a half period of a sinusoid as a lookup function (to compensate for the frequency doubling because of the

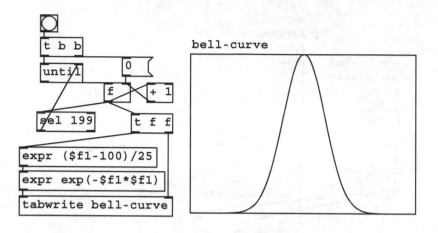

Figure 6.18: Filling in the wavetable for Figure 6.17.

symmetry of the lookup function). To get a half-cycle of the sine function
we multiply the phase by 0.5 and subtract 0.25, so that the adjusted phase
runs from -0.25 to +0.25, once each period. This scans the positive half of
the cycle defined by the cos~ object.

The amplitude of the half-sinusoid is then adjusted by an index of modulation (which is just the bandwidth quotient ω_b/ω). The table ("bell-
curve") holds an unnormalized Gaussian curve sampled from -4 to 4 over
200 points (25 points per unit), so the center of the table, at point 100,
corresponds to the central peak of the bell curve. Outside the interval from
-4 to 4 the Gaussian curve is negligibly small.

Figure 6.18 shows how the Gaussian wavetable is prepared. One new
control object is needed:

until : When the left, "start" inlet is banged, output sequential bangs
(with no elapsed time between them) iteratively, until the right, "stop" inlet
is banged. The stopping "bang" message must originate somehow from
the until object's outlet; otherwise, the outlet will send "bang" messages
forever, freezing out any other object which could break the loop.

As used here, a loop driven by an until object counts from 0 to 199,
inclusive. The loop count is maintained by the "f" and "+ 1" objects, each
of which feeds the other. But since the "+ 1" object's output goes to the
right inlet of the "f", its result (one greater) will only emerge from the
"f" the next time it is banged by "until". So each bang from "until"
increments the value by one.

The order in which the loop is started matters: the upper "t b b" object (short for "trigger bang bang") must first send zero to the "f", thus initializing it, and then set the until object sending bangs, incrementing the value, until stopped. To stop it when the value reaches 199, a select object checks the value and, when it sees the match, bangs the "stop" inlet of the until object.

Meanwhile, for every number from 0 to 199 that comes out of the "f" object, we create an ordered pair of messages to the tabwrite object. First, at right, goes the index itself, from 0 to 199. Then for the left inlet, the first expr object adjusts the index to range from -4 to 4 (it previously ranged from 0 to 199) and the second one evaluates the Gaussian function.

In this patch we have not fully addressed the issue of updating the center frequency quotient at the appropriate times. Whenever the carrier frequency is changed the sample-and-hold step properly delays the update of the quotient. But if, instead or in addition, the fundamental itself changes abruptly, then for a fraction of a period the phasor~ object's frequency and the quotient are out of sync. Pd does not allow the samphold~ output to be connected back into the phasor~ input without the inclusion of an explicit delay (see the next chapter) and there is no simple way to modify the patch to solve this problem.

Assuming that we *did* somehow clock the phasor~ object's input synchronously with its own wraparound points, we would then have to do the same for the bandwidth/fundamental quotient on the right side of the patch as well. In the current scenario, however, there is no problem updating that value continuously.

A practical solution to this updating problem could be simply to rewrite the entire patch in C as a Pd class; this also turns out to use much less CPU time than the pictured patch, and is the more practical solution overall—as long as you don't want to experiment with making embellishments or other changes to the algorithm. Such embellishments might include: adding an inharmonic upward or downward shift in the partials; allowing to switch between smooth and sampled-and-held center frequency updates; adding separate gain controls for even and odd partials; introducing gravel by irregularly modulating the phase; allowing mixtures of two or more waveshaping functions; or making sharper percussive attacks by aligning the phase of the oscillator with the timing of an amplitude envelope generator.

One final detail about amplitude is in order: since the amplitude of the strongest partial decreases roughly as $1/(1 + b)$ where b is the index of modulation, it is sometimes (but not always) desirable to correct the amplitude of the output by multiplying by $1 + b$. This is only an option if b is smoothly updated (as in this example), not if it is sampled-and-held. One situation in which this is appropriate is in simulating plucked strings (by setting center frequency to the fundamental, starting with a high index of modulation and dropping it exponentially); it would be appropriate to hear the fundamental dropping, not rising, in amplitude as the string decays.

Stretched wavetables

Instead of using waveshaping, fomant synthesis is also possible using stretched wavetables, as demonstrated Example F14.wave.packet.pd (not shown here). The technique is essentially that of Example B10.sampler.overlap.pd (described in Section 2.6), with a cosine lookup instead of the more general wavetable, but with the addition of a control to set the duty cycle of the amplitude envelopes. The units are adjusted to be compatible with those of the previous example.

Exercises

1. A pulse train consists of Hann windows (raised cosines), end to end, without any gaps between them. What is the resulting spectrum?

2. To synthesize a formant at 2000 Hertz center frequency and fundamental 300 Hertz, what should the values of k and q be (in the terminology of Figure 6.8)?

3. How would you modify the block diagram of Figure 6.8 to produce only odd harmonics?

Chapter 7

Time Shifts and Delays

At 5:00 some afternoon, put on your favorite recording of the Ramones string quarter number 5. The next Saturday, play the same recording at 5:00:01, one second later in the day. The two playings ideally should sound the same. Shifting the whole thing one second (or, if you like, a few days and a second) has no physical effect on the sound.

But now suppose you played it at 5:00 and 5:00:01 on the same day (on two different playback systems, since the music lasts much longer than one second). Now the sound is much different. The difference, whatever it is, clearly resides in neither of the two individual sounds, but rather in the *interference* between the two. This interference can be perceived in at least four different ways:

Canons: Combining two copies of a signal with a time shift sufficient for the signal to change appreciably, we might hear the two as separate musical streams, in effect comparing the signal to its earlier self. If the signal is a melody, the time shift might be comparable to the length of one or several notes.

Echos: At time shifts between about 30 milliseconds and about a second, the later copy of the signal can sound like an echo of the earlier one. An echo may reduce the intelligibility of the signal (especially if it consists of speech), but usually won't change the overall "shape" of melodies or phrases.

Filtering: At time shifts below about 30 milliseconds, the copies are too close together in time to be perceived separately, and the dominant effect is that some frequencies are enhanced and others suppressed. This changes the spectral envelope of the sound.

Altered room quality: If the second copy is played more quietly than the
first, and especially if we add many more delayed copies at reduced
amplitudes, the result can mimic the echos that arise in a room or
other acoustic space.

The sound of a given arrangement of delayed copies of a signal may combine
two or more of these affects.

Mathematically, the effect of a time shift on a signal can be described
as a phase change of each of the signal's sinusoidal components. The phase
shift of each component is different depending on its frequency (as well as on
the amount of time shift). In the rest of this chapter we will often consider
superpositions of sinusoids at different phases. Heretofore we have been
content to use real-valued sinusoids in our analyses, but in this and later
chapters the formulas will become more complicated and we will need more
powerful mathematical tools to manage them. In a preliminary section of
this chapter we will develop the additional background needed.

7.1   Complex Numbers

Complex numbers are written as:

$$Z = a + bi$$

where a and b are real numbers and $i = \sqrt{-1}$. (In this book we'll use
the upper case Roman letters such as Z to denote complex numbers. Real
numbers appear as lower case Roman or Greek letters, except for integer
bounds, usually written as M or N.) Since a complex number has two real
components, we use a Cartesian plane (in place of a number line) to graph
it, as shown in Figure 7.1. The quantities a and b are called the *real* and
imaginary parts of Z, written as:

$$a = \operatorname{re}(Z)$$

$$b = \operatorname{im}(Z)$$

If Z is a complex number, its *magnitude* (or *absolute value*), written as
$|Z|$, is just the distance in the plane from the origin to the point (a, b):

$$|Z| = \sqrt{(a^2 + b^2)}$$

and its *argument*, written as $\angle(Z)$, is the angle from the positive a axis to
the point (a, b):

$$\angle(Z) = \arctan\left(\frac{b}{a}\right)$$

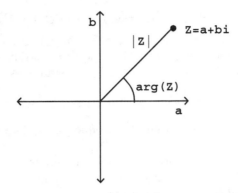

Figure 7.1: A number, Z, in the complex plane. The axes are for the real part a and the imaginary part b.

If we know the magnitude and argument of a complex number (call them r and θ) we can reconstruct the real and imaginary parts:

$$a = r \cos(\theta)$$

$$b = r \sin(\theta)$$

A complex number may be written in terms of its real and imaginary parts a and b, as $Z = a + bi$ (this is called *rectangular form*), or alternatively in *polar form*, in terms of r and θ:

$$Z = r \cdot [\cos(\theta) + i \sin(\theta)]$$

The rectangular and polar formulations are interchangeable; the equations above show how to compute a and b from r and θ and vice versa.

The main reason we use complex numbers in electronic music is because they magically automate trigonometric calculations. We frequently have to add angles together in order to talk about the changing phase of an audio signal as time progresses (or as it is shifted in time, as in this chapter). It turns out that, if you multiply two complex numbers, the argument of the product is the sum of the arguments of the two factors. To see how this happens, we'll multiply two numbers Z_1 and Z_2, written in polar form:

$$Z_1 = r_1 \cdot [\cos(\theta_1) + i \sin(\theta_1)]$$

$$Z_2 = r_2 \cdot [\cos(\theta_2) + i \sin(\theta_2)]$$

giving:

$$Z_1 Z_2 = r_1 r_2 \cdot [\cos(\theta_1) \cos(\theta_2) - \sin(\theta_1) \sin(\theta_2) +$$

$$+i\left(\sin(\theta_1)\cos(\theta_2) + \cos(\theta_1)\sin(\theta_2)\right)]$$

Here the minus sign in front of the $\sin(\theta_1)\sin(\theta_2)$ term comes from multiplying i by itself, which gives -1. We can spot the cosine and sine summation formulas in the above expression, and so it simplifies to:

$$Z_1 Z_2 = r_1 r_2 \cdot \left[\cos(\theta_1 + \theta_2) + i\sin(\theta_1 + \theta_2)\right]$$

By inspection, it follows that the product $Z_1 Z_2$ has magnitude $r_1 r_2$ and argument $\theta_1 + \theta_2$.

We can use this property of complex numbers to add and subtract angles (by multiplying and dividing complex numbers with the appropriate arguments) and then to take the cosine and sine of the result by extracting the real and imaginary parts.

7.1.1 Complex sinusoids

Recall the formula for a (real-valued) sinusoid from Page 1:

$$x[n] = a\cos(\omega n + \phi)$$

This is a sequence of cosines of angles (called phases) which increase arithmetically with the sample number n. The cosines are all adjusted by the factor a. We can now rewrite this as the real part of a much simpler and easier to manipulate sequence of complex numbers, by using the properties of their arguments and magnitudes.

Suppose that a complex number Z happens to have magnitude one and argument ω, so that it can be written as:

$$Z = \cos(\omega) + i\sin(\omega)$$

Then for any integer n, the number Z^n must have magnitude one as well (because magnitudes multiply) and argument $n\omega$ (because arguments add). So,

$$Z^n = \cos(n\omega) + i\sin(n\omega)$$

This is also true for negative values of n, so for example,

$$\frac{1}{Z} = Z^{-1} = cos(\omega) - i\sin(\omega)$$

Figure 7.2 shows graphically how the powers of Z wrap around the unit circle, which is the set of all complex numbers of magnitude one. They form a geometric sequence:

$$\ldots, Z^0, Z^1, Z^2, \ldots$$

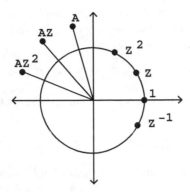

Figure 7.2: The powers of a complex number Z with $|Z| = 1$, and the same sequence multiplied by a constant A.

and taking the real part of each term we get a real sinusoid with initial phase zero and amplitude one:

$$\ldots, \cos(0), \cos(\omega), \cos(2\omega), \ldots$$

Furthermore, suppose we multiply the elements of the sequence by some (complex) constant A with magnitude a and argument ϕ. This gives

$$\ldots, A, AZ, AZ^2, \ldots$$

The magnitudes are all a and the argument of the nth term is $n\omega + \phi$, so the sequence is equal to

$$AZ^n = a \cdot [\cos(n\omega + \phi) + i\sin(n\omega + \phi)]$$

and the real part is just the real-valued sinusoid:

$$\mathrm{re}(AZ^n) = a \cdot \cos(n\omega + \phi)$$

The complex number A encodes both the real amplitude a and the initial phase ϕ; the unit-magnitude complex number Z controls the frequency which is just its argument ω.

Figure 7.2 also shows the sequence $A, AZ, AZ^2, \ldots$; in effect this is the same sequence as $1, Z, Z^2, \ldots$, but amplified and rotated according to the amplitude and initial phase. In a complex sinusoid of this form, A is called the *complex amplitude*.

Using complex numbers to represent the amplitudes and phases of sinusoids can clarify manipulations that otherwise might seem unmotivated.

For instance, suppose we want to know the amplitude and phase of the sum of two sinusoids with the same frequency. In the language of this chapter, we let the two sinusoids be written as:

$$X[n] = AZ^n, \ Y[n] = BZ^n$$

where A and B encode the phases and amplitudes of the two signals. The sum is then equal to:

$$X[n] + Y[n] = (A + B)Z^n$$

which is a sinusoid whose amplitude equals $|A + B|$ and whose phase equals $\angle(A + B)$. This is clearly a much easier way to manipulate amplitudes and phases than using properties of sines and cosines. Eventually, of course, we will take the real part of the result; this can usually be left to the end of whatever we're doing.

7.2 Time Shifts and Phase Changes

Starting from any (real or complex) signal $X[n]$, we can make other signals by time shifting the signal X by a (positive or negative) integer d:

$$Y[n] = X[n - d]$$

so that the dth sample of Y is the 0th sample of X and so on. If the integer d is positive, then Y is a delayed copy of X. If d is negative, then Y anticipates X; this can be done to a recorded sound but isn't practical as a real-time operation.

Time shifting is a linear operation (considered as a function of the input signal X); if you time shift a sum $X_1 + X_2$ you get the same result as if you time shift them separately and add afterward.

Time shifting has the further property that, if you time shift a sinusoid of frequency ω, the result is another sinusoid of the same frequency; time shifting never introduces frequencies that weren't present in the signal before it was shifted. This property, called *time invariance*, makes it easy to analyze the effects of time shifts—and linear combinations of them—by considering separately what the operations do on individual sinusoids.

Furthermore, the effect of a time shift on a sinusoid is simple: it just changes the phase. If we use a complex sinusoid, the effect is even simpler. If for instance

$$X[n] = AZ^n$$

then

$$Y[n] = X[n - d] = AZ^{(n-d)} = Z^{-d}AZ^n = Z^{-d}X[n]$$

so time shifting a complex sinusoid by d samples is the same thing as scaling it by Z^{-d}—it's just an amplitude change by a particular complex number. Since $|Z| = 1$ for a sinusoid, the amplitude change does not change the magnitude of the sinusoid, only its phase.

The phase change is equal to $-d\omega$, where $\omega = \angle(Z)$ is the angular frequency of the sinusoid. This is exactly what we should expect since the sinusoid advances ω radians per sample and it is offset (i.e., delayed) by d samples.

7.3 Delay Networks

If we consider our digital audio samples $X[n]$ to correspond to successive moments in time, then time shifting the signal by d samples corresponds to a *delay* of d/R time units, where R is the sample rate. Figure 7.3 shows one example of a *linear delay network*: an assembly of delay units, possibly with amplitude scaling operations, combined using addition and subtraction. The output is a linear function of the input, in the sense that adding two signals at the input is the same as processing each one separately and adding the results. Moreover, linear delay networks create no new frequencies in the output that weren't present in the input, as long as the network remains time invariant, so that the gains and delay times do not change with time.

In general there are two ways of thinking about delay networks. We can think in the *time domain*, in which we draw waveforms as functions of

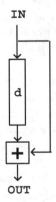

Figure 7.3: A delay network. Here we add the incoming signal to a delayed copy of itself.

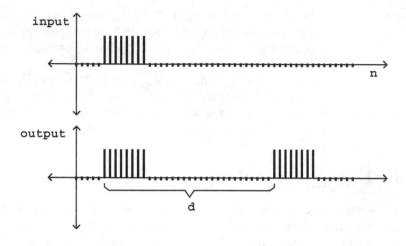

Figure 7.4: The time domain view of the delay network of Figure 7.3. The output is the sum of the input and its time shifted copy.

time (or of the index n), and consider delays as time shifts. Alternatively we may think in the *frequency domain*, in which we dose the input with a complex sinusoid (so that its output is a sinusoid at the same frequency) and report the amplitude and/or phase change wrought by the network, as a function of the frequency. We'll now look at the delay network of Figure 7.3 in each of the two ways in turn.

Figure 7.4 shows the network's behavior in the time domain. We invent some sort of suitable test function as input (it's a rectangular pulse eight samples wide in this example) and graph the input and output as functions of the sample number n. This particular delay network adds the input to a delayed copy of itself.

A frequently used test function is an *impulse*, which is a pulse lasting only one sample. The utility of this is that, if we know the output of the network for an impulse, we can find the output for any other digital audio signal—because any signal $x[n]$ is a sum of impulses, one of height $x[0]$, the next one occurring one sample later and having height $x[1]$, and so on. Later, when the networks get more complicated, we will move to using impulses as input signals to show their time-domain behavior.

On the other hand, we can analyze the same network in the frequency domain by considering a (complex-valued) test signal,

$$X[n] = Z^n$$

where Z has unit magnitude and argument ω. We already know that the output is another complex sinusoid with the same frequency, that is,

$$HZ^N$$

for some complex number H (which we want to find). So we write the output directly as the sum of the input and its delayed copy:

$$Z^n + Z^{-d}Z^n = (1 + Z^{-d})Z^n$$

and find by inspection that:

$$H = 1 + Z^{-d}$$

We can understand the frequency-domain behavior of this delay network by studying how the complex number H varies as a function of the angluar frequency ω. We are especially interested in its argument and magnitude—which tell us the relative phase and amplitude of the sinusoid that comes out. We will work this example out in detail to show how the arithmetic of complex numbers can predict what happens when sinusoids are combined additively.

Figure 7.5 shows the result, in the complex plane, when the quantities 1 and Z^{-d} are combined additively. To add complex numbers we add their real and complex parts separately. So the complex number 1 (real part 1, imaginary part 0) is added coordinate-wise to the complex number Z^{-d} (real part $\cos(-d\omega)$, imaginary part $\sin(-d\omega)$). This is shown graphically by making a parallelogram, with corners at the origin and at the two points to be added, and whose fourth corner is the sum H.

As the figure shows, the result can be understood by symmetrizing it about the real axis: instead of 1 and Z^{-d}, it's easier to sum the quantities $Z^{d/2}$ and $Z^{-d/2}$, because they are symmetric about the real (horizontal) axis. (Strictly speaking, we haven't properly defined the quantities $Z^{d/2}$ and $Z^{-d/2}$; we are using those expressions to denote unit complex numbers whose arguments are half those of Z^d and Z^{-d}, so that squaring them would give Z^d and Z^{-d}.) We rewrite the gain as:

$$H = Z^{-d/2}(Z^{d/2} + Z^{-d/2})$$

The first term is a phase shift of $-d\omega/2$. The second term is best understood in rectangular form:

$$Z^{d/2} + Z^{-d/2}$$

$$= (\cos(\omega d/2) + i\sin(\omega d/2)) + (\cos(\omega d/2) - i\sin(\omega d/2))$$

$$= 2\cos(\omega d/2)$$

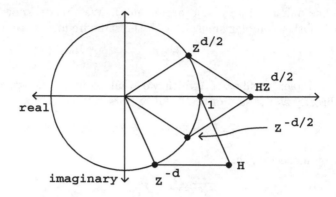

Figure 7.5: Analysis, in the complex plane, of the frequency-domain be-havior of the delay network of Figure 7.3. The complex number Z encodes the frequency of the input. The delay line output is the input times Z^{-d}. The total (complex) gain is H. We find the magnitude and argument of H by symmetrizing the sum, rotating it by $d/2$ times the angular frequency of the input.

This real-valued quantity may be either positive or negative; its absolute value gives the magnitude of the output:

$$|H| = 2|\cos(\omega d/2)|$$

The quantity $|H|$ is called the *gain* of the delay network at the angular frequency ω, and is graphed in Figure 7.6. The frequency-dependent gain of a delay network (that is, the gain as a function of frequency) is called the network's *frequency response.*

Since the network has greater gain at some frequencies than at others, it may be considered as a *filter* that can be used to separate certain com-ponents of a sound from others. Because of the shape of this particular gain expression as a function of ω, this kind of delay network is called a (non-recirculating) *comb filter.*

The output of the network is a sum of two sinusoids of equal amplitude, and whose phases differ by ωd. The resulting frequency response agrees with common sense: if the angular frequency ω is set so that an integer number of periods fit into d samples, i.e., if ω is a multiple of $2\pi/d$, the output of the delay is exactly the same as the original signal, and so the two combine to make an output with twice the original amplitude. On

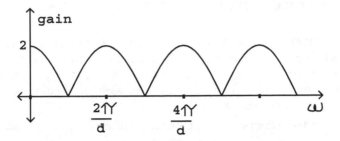

Figure 7.6: Gain of the delay network of Figure 7.3, shown as a function of angular frequency ω.

the other hand, if for example we take $\omega = \pi/d$ so that the delay is half the period, then the delay output is out of phase and cancels the input exactly.

This particular delay network has an interesting application: if we have a periodic (or nearly periodic) incoming signal, whose fundamental frequency is ω radians per sample, we can tune the comb filter so that the peaks in the gain are aligned at even harmonics and the odd ones fall where the gain is zero. To do this we choose $d = \pi/\omega$, i.e., set the delay time to exactly one half period of the incoming signal. In this way we get a new signal whose harmonics are $2\omega, 4\omega, 6\omega, \ldots$, and so it now has a new fundamental frequency at twice the original one. Except for a factor of two, the amplitudes of the remaining harmonics still follow the spectral envelope of the original sound. So we have a tool now for raising the pitch of an incoming sound by an octave without changing its spectral envelope. This octave doubler is the reverse of the octave divider introduced back in Chapter 5.

The time and frequency domains offer complementary ways of looking at the same delay network. When the delays inside the network are smaller than the ear's ability to resolve events in time—less than about 20 milliseconds—the time domain picture becomes less relevant to our understanding of the delay network, and we turn mostly to the frequency-domain picture. On the other hand, when delays are greater than about 50 msec, the peaks and valleys of plots showing gain versus frequency (such as that of Figure 7.6) crowd so closely together that the frequency-domain view becomes less important. Both are nonetheless valid over the entire range of possible delay times.

7.4   Recirculating Delay Networks

It is sometimes desirable to connect the outputs of one or more delays in a network back into their own or each others' inputs. Instead of getting one or several echos of the original sound as in the example above, we can potentially get an infinite number of echos, each one feeding back into the network to engender yet others.

The simplest example of a recirculating network is the *recirculating comb filter* whose block diagram is shown in Figure 7.7. As with the earlier, simple comb filter, the input signal is sent down a delay line whose length is d samples. But now the delay line's output is also fed back to its input; the delay's input is the sum of the original input and the delay's output. The output is multiplied by a number g before feeding it back into its input.

The time domain behavior of the recirculating comb filter is shown in Figure 7.8. Here we consider the effect of sending an impulse into the network. We get back the original impulse, plus a series of echos, each in turn d samples after the previous one, and multiplied each time by the gain g. In general, a delay network's output given an impulse as input is called the network's *impulse response*.

Note that we have chosen a gain g that is less than one in absolute value. If we chose a gain greater than one (or less than -1), each echo would have a larger magnitude than the previous one. Instead of falling exponentially as they do in the figure, they would grow exponentially. A recirculating network whose output eventually falls toward zero after its

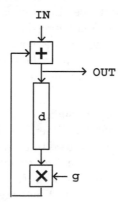

Figure 7.7: Block diagram for a recirculating comb filter. Here d is the delay time in samples and g is the feedback coefficient.

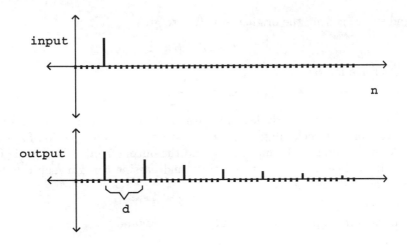

Figure 7.8: Time-domain analysis of the recirculating comb filter, using an impulse as input.

input terminates is called *stable*; one whose output grows without bound is called *unstable*.

We can also analyse the recirculating comb filter in the frequency domain. The situation is now quite hard to analyze using real sinusoids, and so we get the first big payoff for having introduced complex numbers, which greatly simplify the analysis.

If, as before, we feed the input with the signal,

$$X[n] = Z^n$$

with $|Z| = 1$, we can write the output as

$$Y[n] = (1 + gZ^{-d} + g^2 Z^{-2d} + \cdots)X[n]$$

Here the terms in the sum come from the series of discrete echos. It follows that the amplitude of the output is:

$$H = 1 + gZ^{-d} + (gZ^{-d})^2 + \cdots$$

This is a geometric series; we can sum it using the standard technique. First multiply both sides by gZ^{-d} to give:

$$gZ^{-d}H = gZ^{-d} + (gZ^{-d})^2 + (gZ^{-d})^3 + \cdots$$

and subtract from the original equation to give:

$$H - gZ^{-d}H = 1$$

Then solve for H:

$$H = \frac{1}{1 - gZ^{-d}}$$

A faster (but slightly less intuitive) method to get the same result is to examine the recirculating network itself to yield an equation for H, as follows. We named the input $X[n]$ and the output $Y[n]$. The signal going into the delay line is the output $Y[n]$, and passing this through the delay line and multiplier gives

$$Y[n] \cdot gZ^{-d}$$

This plus the input is just the output signal again, so:

$$Y[n] = X[n] + Y[n] \cdot gZ^{-d}$$

and dividing by $X[n]$ and using $H = Y[n]/X[n]$ gives:

$$H = 1 + HgZ^{-d}$$

This is equivalent to the earlier equation for H.

Now we would like to make a graph of the frequency response (the gain as a function of frequency) as we did for non-recirculating comb filters in Figure 7.6. This again requires that we make a preliminary picture in the complex plane. We would like to estimate the magnitude of H equal to:

$$|H| = \frac{1}{|1 - gZ^{-d}|}$$

where we used the multiplicative property of magnitudes to conclude that the magnitude of a (complex) reciprocal is the reciprocal of a (real) magnitude. Figure 7.9 shows the situation graphically. The gain $|H|$ is the reciprocal of the length of the segment reaching from the point 1 to the point gZ^{-d}. Figure 7.10 shows a graph of the frequency response $|H|$ as a function of the angular frequency $\omega = \angle(Z)$.

Figure 7.9 can be used to analyze how the frequency response $|H(\omega)|$ should behave qualitatively as a function of g. The height and bandwidth of the peaks both depend on g. The maximum value that $|H|$ can attain is when

$$Z^{-d} = 1$$

This occurs at the frequencies $\omega = 0, 2\pi/d, 4\pi/d, \ldots$ as in the simple comb filter above. At these frequencies the gain reaches

$$|H| = \frac{1}{1 - g}$$

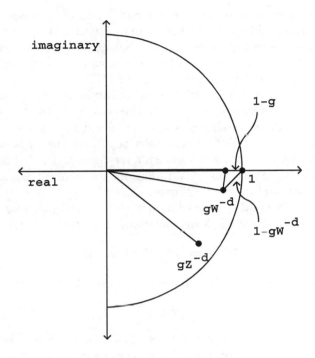

Figure 7.9: Diagram in the complex plane for approximating the output gain $|H|$ of the recirculating comb filters at three different frequencies: 0, and the arguments of two unit complex numbers W and Z; W is chosen to give a gain about 3 dB below the peak.

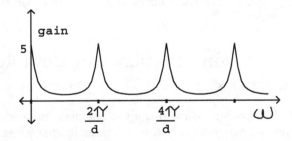

Figure 7.10: Frequency response of the recirculating comb filter with $g = 0.8$. The peak gain is $1/(1-g) = 5$. Peaks are much narrower than for the non-recirculating comb filter.

The next important question is the bandwidth of the peaks in the frequency response. So we would like to find sinusoids W^n, with frequency $\angle(W)$, giving rise to a value of $|H|$ that is, say, 3 decibels below the maximum. To do this, we return to Figure 7.9, and try to place W so that the distance from the point 1 to the point gW^{-d} is about $\sqrt{2}$ times the distance from 1 to g (since $\sqrt{2}$:1 is a ratio of approximately 3 decibels).

We do this by arranging for the imaginary part of gW^{-d} to be roughly $1 - g$ or its negative, making a nearly isosceles right triangle between the points 1, $1 - g$, and gW^{-d}. (Here we're supposing that g is at least 2/3 or so; otherwise this approximation isn't very good). The hypotenuse of a right isosceles triangle is always $\sqrt{2}$ times the leg, and so the gain drops by that factor compared to its maximum.

We now make another approximation, that the imaginary part of gW^{-d} is approximately the angle in radians it cuts from the real axis:

$$\pm(1 - g) \approx \text{im}(gW^{-d}) \approx \angle(W^{-d})$$

So the region of each peak reaching within 3 decibels of the maximum value is about

$$(1 - g)/d$$

(in radians) to either side of the peak. The bandwidth narrows (and the filter peaks become sharper) as g approaches its maximum value of 1.

As with the non-recirculating comb filter of Section 7.3, the teeth of the comb are closer together for larger values of the delay d. On the other hand, a delay of $d = 1$ (the shortest possible) gets only one tooth (at zero frequency) below the Nyquist frequency π (the next tooth, at 2π, corresponds again to a frequency of zero by foldover). So the recirculating comb filter with $d = 1$ is just a low-pass filter. Delay networks with one-sample delays will be the basis for designing many other kinds of digital filters in Chapter 8.

7.5   Power Conservation and Complex Delay Networks

The same techniques will work to analyze any delay network, although for more complicated networks it becomes harder to characterize the results, or to design the network to have specific, desired properties. Another point of view can sometimes be usefully brought to the situation, particularly when flat frequency responses are needed, either in their own right or else to ensure that a complex, recirculating network remains stable at feedback gains close to one.

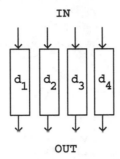

Figure 7.11: First fundamental building block for unitary delay networks: delay lines in parallel.

The central fact we will use is that if any delay network, with either one or many inputs and outputs, is constructed so that its output power (averaged over time) always equals its input power, that network has to have a flat frequency response. This is almost a tautology; if you put in a sinusoid at any frequency on one of the inputs, you will get sinusoids of the same frequency at the outputs, and the sum of the power on all the outputs will equal the power of the input, so the gain, suitably defined, is exactly one.

In order to work with power-conserving delay networks we will need an explicit definition of "total average power". If there is only one signal (call it $x[n]$), the average power is given by:

$$P(x[n]) = \left[|x[0]|^2 + |x[1]|^2 + \cdots + |x[N-1]|^2 \right] / N$$

where N is a large enough number so that any fluctuations in amplitude get averaged out. This definition works as well for complex-valued signals as for real-valued ones. The average total power for several digital audio signals is just the sum of the individual signal's powers:

$$P(x_1[n], \ldots, x_r[n]) = P(x_1[n]) + \cdots + P(x_r[n])$$

where r is the number of signals to be combined.

It turns out that a wide range of interesting delay networks has the property that the total power output equals the total power input; they are called *unitary*. To start with, we can put any number of delays in parallel, as shown in Figure 7.11. Whatever the total power of the inputs, the total power of the outputs has to equal it.

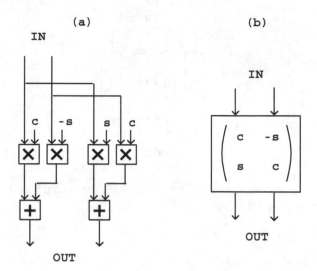

Figure 7.12: Second fundamental building block for unitary delay networks: rotating two digital audio signals. Part (a) shows the transformation explicitly; (b) shows it as a matrix operation.

A second family of power-preserving transformations is composed of rotations and reflections of the signals $x_1[n]$, ... , $x_r[n]$, considering them, at each fixed time point n, as the r coordinates of a point in r-dimensional space. The rotation or reflection must be one that leaves the origin $(0, \ldots, 0)$ fixed.

For each sample number n, the total contribution to the average signal power is proportional to

$$|x_1|^2 + \cdots + |x_r|^2$$

This is just the Pythagorean distance from the origin to the point $(x_1, \ldots, x_r)$. Since rotations and reflections are distance-preserving transformations, the distance from the origin before transforming must equal the distance from the origin afterward. So the total power of a collection of signals must must be preserved by rotation.

Figure 7.12 shows a rotation matrix operating on two signals. In part (a) the transformation is shown explicitly. If the input signals are $x_1[n]$ and $x_2[n]$, the outputs are:

$$y_1[n] = cx_1[n] - sx_2[n]$$

$$y_2[n] = sx_1[n] + cx_2[n]$$

where c, s are given by

$$c = \cos(\theta)$$

$$s = \sin(\theta)$$

for an *angle of rotation* θ. Considered as points on the Cartesian plane, the point (y_1, y_2) is just the point (x_1, x_2) rotated counter-clockwise by the angle θ. The two points are thus at the same distance from the origin:

$$|y_1|^2 + |y_2|^2 = |x_1|^2 + |x_2|^2$$

and so the two output signals have the same total power as the two input signals.

For an alternative description of rotation in two dimensions, consider complex numbers $X = x_1 + x_2 i$ and $Y = y_1 + y_2 i$. The above transformation amounts to setting

$$Y = XZ$$

where Z is a complex number with unit magnitude and argument θ. Since $|Z| = 1$, it follows that $|X| = |Y|$.

If we perform a rotation on a pair of signals and then invert one (but not the other) of them, the result is a *reflection*. This also preserves total signal power, since we can invert any or all of a collection of signals without changing the total power. In two dimensions, a reflection appears as a transformation of the form

$$y_1[n] = cx_1[n] + sx_2[n]$$

$$y_2[n] = sx_1[n] - cx_2[n]$$

A special and useful rotation matrix is obtained by setting $\theta = \pi/4$, so that $s = c = \sqrt{1/2}$. This allows us to simplify the computation as shown in Figure 7.13 (part a) because each signal need only be multiplied by the one quantity $c = s$.

More complicated rotations or reflections of more than two input signals may be made by repeatedly rotating and/or reflecting them in pairs. For example, in Figure 7.13 (part b), four signals are combined in pairs, in two successive stages, so that in the end every signal input feeds into all the outputs. We could do the same with eight signals (using three stages) and so on. Furthermore, if we use the special angle $\pi/4$, all the input signals will contribute equally to each of the outputs.

Any combination of delays and rotation matrices, applied in succession to a collection of audio signals, will result in a flat frequency response,

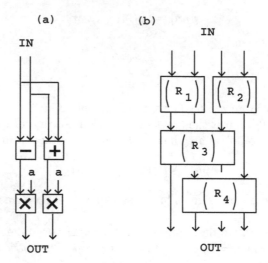

Figure 7.13: Details about rotation (and reflection) matrix operations: (a) rotation by the angle $\theta = \pi/4$, so that $a = \cos(\theta) = \sin(\theta) = \sqrt{1/2} \approx 0.7071$; (b) combining two-dimensional rotations to make higher-dimensional ones.

since each individual operation does. This already allows us to generate an infinitude of flat-response delay networks, but so far, none of them are recirculating. A third operation, shown in Figure 7.14, allows us to make recirculating networks that still enjoy flat frequency responses.

Part (a) of the figure shows the general layout. The transformation R is assumed to be any combination of delays and mixing matrices that preserves total power. The signals $x_1, \ldots x_k$ go into a unitary delay network, and the output signals $y_1, \ldots y_k$ emerge. Some other signals $w_1, \ldots w_j$ (where j is not necessarily equal to k) appear at the output of the transformation R and are fed back to its input.

If R is indeed power preserving, the total input power (the power of the signals $x_1, \ldots x_k$ plus that of the signals $w_1, \ldots w_j$) must equal the output power (the power of the signals $y_1, \ldots y_k$ plus $w_1, \ldots w_j$), and subtracting all the w from the equality, we find that the total input and output power are equal.

If we let $j = k = 1$ so that there is one x, y, and w, and let the transformation R be a rotation by θ followed by a delay of d samples on the W output, the result is the well-known *all-pass filter*. With some juggling, and letting $c = \cos(\theta)$, we can show it is equivalent to the network shown

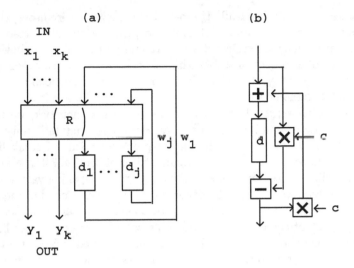

Figure 7.14: Flat frequency response in recirculating networks: (a) in general, using a rotation matrix R; (b) the "all-pass" configuration.

in part (b) of the figure. All-pass filters have many applications, some of which we will visit later in this book.

7.6 Artificial Reverberation

Artificial reverberation is widely used to improve the sound of recordings, but has a wide range of other musical applications [DJ85, pp.289-340]. Reverberation in real, natural spaces arises from a complicated pattern of sound reflections off the walls and other objects that define the space. It is a great oversimplification to imitate this process using recirculating, discrete delay networks. Nonetheless, modeling reverberation using recirculating delay lines can, with much work, be made to yield good results.

The central idea is to idealize any room (or other reverberant space) as a collection of parallel delay lines that models the memory of the air inside the room. At each point on the walls of the room, many straight-line paths terminate, each carrying sound to that point; the sound then reflects into many other paths, each one originating at that point, and leading eventually to some other point on a wall.

Although the wall (and the air we passed through to get to the wall) absorbs some of the sound, some portion of the incident power is reflected and makes it to another wall. If most of the energy recirculates, the room reverberates for a long time; if all of it does, the reverberation lasts forever. If at any frequency the walls reflect more energy overall than they receive, the sound will feed back unstably; this never happens in real rooms (conservation of energy prevents it), but it can happen in an artificial reverberator if it is not designed correctly.

To make an artificial reverberator using a delay network, we must fill two competing demands simultaneously. First, the delay lines must be long enough to prevent *coloration* in the output as a result of comb filtering. (Even if we move beyond the simple comb filter of Section 7.4, the frequency response will tend to have peaks and valleys whose spacing varies inversely with total delay time.) On the other hand, we should not hear individual echoes; the *echo density* should ideally be at least one thousand per second.

In pursuit of these aims, we assemble some number of delay lines and connect their outputs back to their inputs. The feedback path—the connection from the outputs back to the inputs of the delays—should have an aggregate gain that varies gently as a function of frequency, and never exceeds one for any frequency. A good starting point is to give the feedback path a flat frequency response and a gain slightly less than one; this is done using rotation matrices.

Ideally this is all we should need to do, but in reality we will not always want to use the thousands of delay lines it would take to model the paths between every possible pair of points on the walls. In practice we usually use between four and sixteen delay lines to model the room. This simplification sometimes reduces the echo density below what we would wish, so we might use more delay lines at the input of the recirculating network to increase the density.

Figure 7.15 shows a simple reverberator design that uses this principle. The incoming sound, shown as two separate signals in this example, is first thickened by progressively delaying one of the two signals and then intermixing them using a rotation matrix. At each stage the number of echoes of the original signal is doubled; typically we would use between 6 and 8 stages to make between 64 and 256 echos, all with a total delay of between 30 and 80 milliseconds. The figure shows three such stages.

Next comes the recirculating part of the reverberator. After the initial thickening, the input signal is fed into a bank of parallel delay lines, and their outputs are again mixed using a rotation matrix. The mixed outputs are attenuated by a gain $g \leq 1$, and fed back into the delay lines to make a recirculating network.

The value g controls the reverberation time. If the average length of the recirculating delay lines is d, then any incoming sound is attenuated by a factor of g after a time delay of d. After time t the signal has recirculated

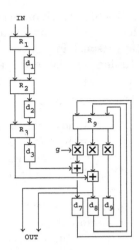

Figure 7.15: Reverberator design using power-preserving transformations and recirculating delays.

t/d times, losing $20\log_{10}(g)$ decibels each time around, so the total gain, in decibels, is:

$$20\frac{t}{d}\log_{10}(g)$$

The usual measure of reverberation time (RT) is the time at which the gain drops by sixty decibels:

$$20\frac{\text{RT}}{d}\log_{10}(g) = -60$$

$$\text{RT} = \frac{-3d}{\log_{10}(g)}$$

If g is one, this formula gives ∞, since the logarithm of one is zero.

The framework shown above is the basis for many modern reverberator designs. Many extensions of this underlying design have been proposed. The most important next step would be to introduce filters in the recirculation path so that high frequencies can be made to decay more rapidly than low ones; this is readily accomplished with a very simple low-pass filter, but we will not work this out here, having not yet developed the needed filter theory.

In general, to use this framework to design a reverberator involves making many complicated choices of delay times, gains, and filter coefficients. Mountains of literature have been published on this topic; Barry Blesser has published a good overview [Ble01]. Much more is known about reverberator design and tuning that has not been published; precise designs are often kept secret for commercial reasons. In general, the design process involves painstaking and lengthy tuning by trial, error, and critical listening.

7.6.1   Controlling reverberators

Artificial reverberation is used almost universally in recording or sound reinforcement to sweeten the overall sound. However, and more interestingly, reverberation may be used as a sound source in its own right. The special case of infinite reverberation is useful for grabbing live sounds and extending them in time.

To make this work in practice it is necessary to open the input of the reverberator only for a short period of time, during which the input sound is not varying too rapidly. If an infinite reverberator's input is left open for too long, the sound will collect and quickly become an indecipherable mass. To "infinitely reverberate" a note of a live instrument, it is best to wait until after the attack portion of the note and then allow perhaps 1/2 second of the note's steady state to enter the reverberator. It is possible to build chords from a monophonic instrument by repeatedly opening the input at different moments of stable pitch.

Figure 7.16 shows how this can be done in practice. The two most important controls are the reverberator's input and feedback gains. To capture a sound, we set the feedback gain to one (infinite reverberation time) and momentarily open the input at time t_1. To add other sounds to an already held one, we simply reopen the input gain at the appropriate moments (at time t_2 in the figure, for example). Finally, we can erase the recirculating sound, thus both fading the output and emptying the reverberator, by setting the feedback gain to a value less than one (as at time t_3). The further we reduce the feedback gain, the faster the output will decay.

7.7   Variable and Fractional Shifts

Like any audio synthesis or processing technique, delay networks become much more powerful and interesting if their characteristics can be made to change over time. The gain parameters (such as g in the recirculating comb filter) may be controlled by envelope generators, varying them while avoiding clicks or other artifacts. The delay times (such as d before) are not so easy to vary smoothly for two reasons.

First, we have only defined time shifts for integer values of d, since for fractional values of d an expression such as $x[n-d]$ is not determined if

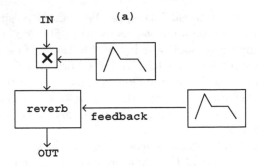

IN (a)

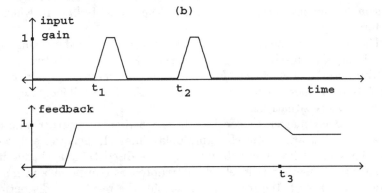

Figure 7.16: Controlling a reverberator to capture sounds selectively: (a) the network; (b) examples of how to control the input gain and feedback to capture two sounds at times t_1 and t_2, and to hold them until a later time t_3.

$x[n]$ is only defined for integer values of n. To make fractional delays we will have to introduce some suitable interpolation scheme. And if we wish to vary d smoothly over time, it will not give good results simply to hop from one integer to the next.

Second, even once we have achieved perfectly smoothly changing delay times, the artifacts caused by varying delay time become noticeable even at very small relative rates of change; while in most cases you may ramp an amplitude control between any two values over 30 milliseconds without trouble, changing a delay by only one sample out of every hundred makes a very noticeable shift in pitch—indeed, one frequently will vary a delay deliberately in order to hear the artifacts, only incidentally passing from one specific delay time value to another one.

The first matter (fractional delays) can be dealt with using an interpolation scheme, in exactly the same way as for wavetable lookup (Section 2.5). For example, suppose we want a delay of $d = 1.5$ samples. For each n we must estimate a value for $x[n - 1.5]$. We could do this using standard four-point interpolation, putting a cubic polynomial through the four "known" points (0, x[n]), (1, x[n-1]), (2, x[n-2]), (3, x[n-3]), and then evaluating the polynomial at the point 1.5. Doing this repeatedly for each value of n gives the delayed signal.

This four-point interpolation scheme can be used for any delay of at least one sample. Delays of less than one sample can't be calculated this way because we need two input points at least as recent as the desired delay. They were available in the above example, but for a delay time of 0.5 samples, for instance, we would need the value of $x[n + 1]$, which is in the future.

The accuracy of the estimate could be further improved by using higher-order interpolation schemes. However, there is a trade-off between quality and computational efficiency. Furthermore, if we move to higher-order interpolation schemes, the minimum possible delay time will increase, causing trouble in some situations.

The second matter to consider is the artifacts—whether wanted or unwanted—that arise from changing delay lines. In general, a discontinuous change in delay time will give rise to a discontinuous change in the output signal, since it is in effect interrupted at one point and made to jump to another. If the input is a sinusoid, the result is a discontinuous phase change.

If it is desired to change the delay line occasionally between fixed delay times (for instance, at the beginnings of musical notes), then we can use the techniques for managing sporadic discontinuities that were introduced in Section 4.3. In effect these techniques all work by muting the output in one way or another. On the other hand, if it is desired that the delay time change continuously—while we are listening to the output—then we must take into account the artifacts that result from the changes.

Figure 7.17 shows the relationship between input and output time in a variable delay line. The delay line is assumed to have a fixed maximum length D. At each sample of output (corresponding to a point on the horizontal axis), we output one (possibly interpolated) sample of the delay line's input. The vertical axis shows which sample (integer or fractional) to use from the input signal. Letting n denote the output sample number, the vertical axis shows the quantity $n - d[n]$, where $d[n]$ is the (time-varying) delay in samples. If we denote the input sample location by:

$$y[n] = n - d[n]$$

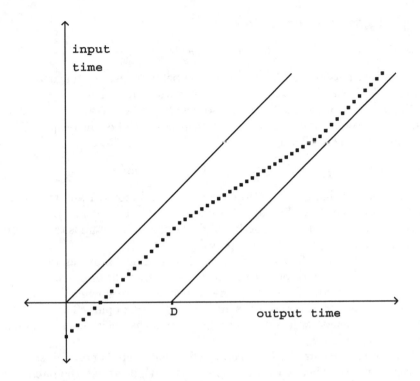

Figure 7.17: A variable length delay line, whose output is the input from some previous time. The output samples can't be newer than the input samples, nor older than the length D of the delay line. The slope of the input/output curve controls the momentary transposition of the output.

then the output of the delay line is:

$$z[n] = x[y[n]]$$

where the signal x is evaluated at the point $y[n]$, interpolating appropriately in case $y[n]$ is not an integer. This is exactly the formula for wavetable lookup (Page 27). We can use all the properties of wavetable lookup of recorded sounds to predict the behavior of variable delay lines.

There remains one difference between delay lines and wavetables: the material in the delay line is constantly being refreshed. Not only can we not read into the future, but, if the the delay line is D samples in length, we can't read further than D samples into the past either:

$$0 < d[n] < D$$

or, negating this and adding n to each side,

$$n > y[n] > n - D.$$

This last relationship appears as the region between the two diagonal lines in Figure 7.17; the function $y[n]$ must stay within this strip.

Returning to Section 2.2, we can use the Momentary Transposition Formulas for wavetables to calculate the transposition $t[n]$ of the output. This gives the Momentary Transposition Formula for delay lines:

$$t[n] = y[n] - y[n-1] = 1 - (d[n] - d[n-1])$$

If $d[n]$ does not change with n, the transposition factor is 1 and the sound emerges from the delay line at the same speed as it went in. But if the delay time is increasing as a function of n, the resulting sound is transposed downward, and if $d[n]$ decreases, upward.

This is called the *Doppler effect*, and it occurs in nature as well. The air that sound travels through can sometimes be thought of as a delay line. Changing the length of the delay line corresponds to moving the listener toward or away from a stationary sound source; the Doppler effect from the changing path length works precisely the same in the delay line as it would be in the physical air.

Returning to Figure 7.17, we can predict that there is no pitch shift at the beginning, but then when the slope of the path decreases the pitch will drop for an interval of time before going back to the original pitch (when the slope returns to one). The delay time can be manipulated to give any desired transposition, but the greater the transposition, the less long we can maintain it before we run off the bottom or the top of the diagonal region.

7.8   Fidelity of Interpolating Delay Lines

Since they are in effect doing wavetable lookup, variable delay lines introduce distortion to the signals they operate on. Moreover, a subtler problem can come up even when the delay line is not changing in length: the frequency response, in real situations, is never perfectly flat for a delay line whose length is not an integer.

If the delay time is changing from sample to sample, the distortion results of Section 2.5 apply. To use them, we suppose that the delay line input can be broken down into sinusoids and consider separately what happens to each individual sinusoid. We can use Table 2.1 (Page 46) to predict the RMS level of the combined distortion products for an interpolated variable delay line.

We'll assume here that we want to use four-point interpolation. For sinusoids with periods longer than 32 samples (that is, for frequencies below 1/16 of the Nyquist frequency) the distortion is 96 dB or better—unlikely ever to be noticeable. At a 44 kHz. sample rate, these periods would correspond to frequencies up to about 1400 Hertz. At higher frequencies the quality degrades, and above 1/4 the Nyquist frequency the distortion products, which are only down about 50 dB, will probably be audible.

The situation for a complex tone depends primarily on the amplitudes and frequencies of its higher partials. Suppose, for instance, that a tone's partials above 5000 Hertz are at least 20 dB less than its strongest partial, and that above 10000 Hertz they are down 60 dB. Then as a rough estimate, the distortion products from the range 5000-10000 will each be limited to about -68 dB and those from above 10000 Hertz will be limited to about -75 dB (because the worst figure in the table is about -15 dB and this must be added to the strength of the partial involved.)

If the high-frequency content of the input signal does turn out to give unacceptable distortion products, in general it is more effective to increase the sample rate than the number of points of interpolation. For periods greater than 4 samples, doubling the period (by doubling the sample rate, for example) decreases distortion by about 24 dB.

The 4-point interpolating delay line's frequency response is nearly flat up to half the Nyquist frequency, but thereafter it dives quickly. Suppose (to pick the worst case) that the delay is set halfway between two integers, say 1.5. Cubic interpolation gives:

$$x[1.5] = \frac{-x[0] + 9x[1] + 9x[2] - x[3]}{8}$$

Now let $x[n]$ be a (real-valued) unit-amplitude sinusoid with angular frequency ω, whose phase is zero at 1.5:

$$x[n] = \cos(\omega \cdot (n - 1.5))$$

and compute $x[1.5]$ using the above formula:

$$x[1.5] = \frac{9\cos(\omega/2) - \cos(3\omega/2)}{4}$$

This is the peak value of the sinusoid that comes back out of the delay line, and since the peak amplitude going in was one, this shows the frequency response of the delay line. This is graphed in Figure 7.18. At half the Nyquist frequency ($\omega = \pi/2$) the gain is about -1 dB, which is a barely perceptible drop in amplitude. At the Nyquist frequency itself, however, the gain is zero.

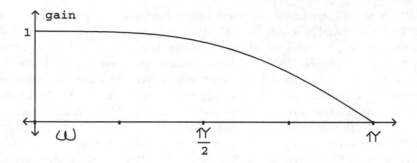

Figure 7.18: Gain of a four-point interpolating delay line with a delay halfway between two integers. The DC gain is one.

As with the results for distortion, the frequency response improves radically with a doubling of sample rate. If we run our delay at a sample rate of 88200 Hertz instead of the standard 44100, we will get only about 1 dB of roll-off all the way up to 20000 Hertz.

7.9 Pitch Shifting

A favorite use of variable delay lines is to alter the pitch of an incoming sound using the Doppler effect. It may be desirable to alter the pitch variably (randomly or periodically, for example), or else to maintain a fixed musical interval of transposition over a length of time.

Returning to Figure 7.17, we see that with a single variable delay line we can maintain any desired pitch shift for a limited interval of time, but if we wish to sustain a fixed transposition we will always eventually land outside the diagonal strip of admissible delay times. In the simplest scenario, we simply vary the transposition up and down so as to remain in the strip.

This works, for example, if we wish to apply vibrato to a sound as shown in Figure 7.19. Here the delay function is

$$d[n] = d_0 + a\cos(\omega n)$$

where d_0 is the average delay, a is the amplitude of variation about the average delay, and ω is an angular frequency. The Momentary Transposition (Page 202), is approximately

$$t = 1 + a\omega\cos(\omega n - \pi/2)$$

This ranges in value between $1 - a\omega$ and $1 + a\omega$.

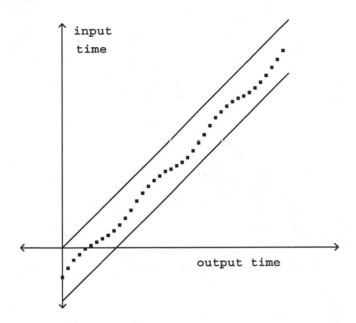

Figure 7.19: Vibrato using a variable delay line. Since the pitch shift alternates between upward and downward, it is possible to maintain it without drifting outside the strip of admissible delay.

Suppose, on the other hand, that we wish to maintain a constant transposition over a longer interval of time. In this case we can't maintain the transposition forever, but it is still possible to maintain it over fixed intervals of time broken by discontinuous changes, as shown in Figure 7.20. The delay time is the output of a suitably normalized sawtooth function, and the output of the variable delay line is enveloped as shown in the figure to avoid discontinuities.

This is accomplished as shown in Figure 7.21. The output of the sawtooth generator is used in two ways. First it is adjusted to run between the bounds d_0 and $d_0 + s$, and this adjusted sawtooth controls the delay time, in samples. The initial delay d_0 should be at least enough to make the variable delay feasible; for four-point interpolation it must be at least one sample. Larger values of d_0 add a constant, additional delay to the output; this is usually offered as a control in a pitch shifter since it is essentially free. The quantity s is sometimes called the *window size*. It corresponds roughly to the sample length in a looping sampler (Section 2.2).

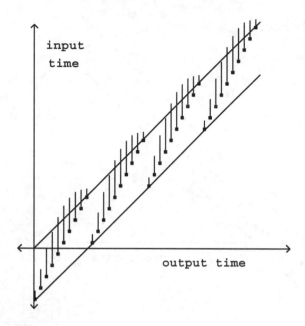

Figure 7.20: Piecewise linear delay functions to maintain a constant transposition (except at the points of discontinuity). The outputs are enveloped as suggested by the bars above each point, to smooth the output at the points of discontinuity in delay time.

The sawtooth output is also used to envelope the output in exactly the same way as in the enveloped wavetable sampler of Figure 2.7 (Page 38). The envelope is zero at the points where the sawtooth wraps around, and in between, rises smoothly to a maximum value of 1 (for unit gain).

If the frequency of the sawtooth wave is f (in cycles per second), then its value sweeps from 0 to 1 every R/f samples (where R is the sample rate). The difference between successive values is thus f/R. If we let $x[n]$ denote the output of the sawtooth oscillator, then

$$x[n+1] - x[n] = \frac{f}{R}$$

(except at the wraparound points). If we adjust the output range of the wavetable oscillator to the value s (as is done in the figure) we get a new slope:

$$s \cdot x[n+1] - s \cdot x[n] = \frac{sf}{R}$$

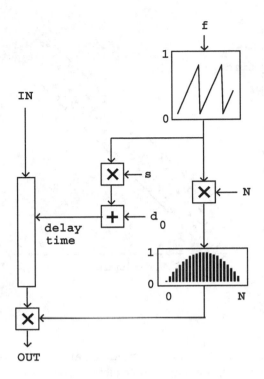

Figure 7.21: Using a variable delay line as a pitch shifter. The sawtooth wave creates a smoothly increasing or decreasing delay time. The output of the delay line is enveloped to avoid discontinuities. Another copy of the same diagram should run 180 degrees (π radians) out of phase with this one.

Adding the constant d_0 has no effect on this slope. The Momentary Transposition (Page 202) is then:

$$t = 1 - \frac{sf}{R}$$

To complete the design of the pitch shifter we must add the other copy halfway out of phase. This gives rise to a delay reading pattern as shown in Figure 7.22.

The pitch shifter can transpose either upward (using negative sawtooth frequencies, as in the figure) or downward, using positive ones. Pitch shift is usually controlled by changing f with s fixed. To get a desired transposition

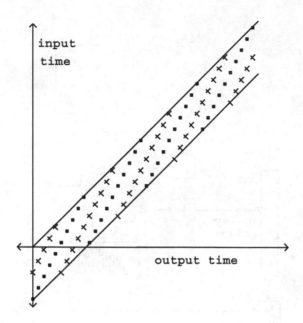

Figure 7.22: The pitch shifter's delay reading pattern using two delay lines, so that one is at maximum amplitude exactly when the other is switching.

interval t, set

$$f = \frac{(t-1)R}{s}$$

The window size s should be chosen small enough, if possible, so that the two delayed copies ($s/2$ samples apart) do not sound as distinct echoes. However, very small values of s will force f upward; values of f greater than about 5 Hertz result in very audible modulation. So if very large transpositions are required, the value of s may need to be increased. Typical values range from 30 to 100 milliseconds (about $R/30$ to $R/10$ samples).

Although the frequency may be changed at will, even discontinuously, s must be changed more carefully. A possible solution is to mute the output while changing s discontinuously; alternatively, s may be ramped continuously but this causes hard-to-control Doppler shifts.

A good choice of envelope is one half cycle of a sinusoid. If we assume on average that the two delay outputs are uncorrelated (Page 12), the signal power from the two delay lines, after enveloping, will add to a constant (since the sum of squares of the two envelopes is one).

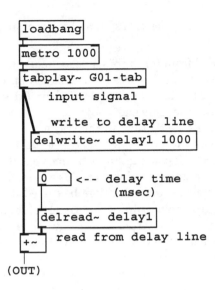

Figure 7.23: Example patch G01.delay.pd, showing a noninterpolating delay with a delay time controlled in milliseconds.

Many variations exist on this pitch shifting algorithm. One classic variant uses a single delay line, with no enveloping at all. In this situation it is necessary to choose the point at which the delay time jumps, and the point it jumps to, so that the output stays continuous. For example, one could find a point where the output signal passes through zero (a "zero crossing") and jump discontinuously to another one. Using only one delay line has the advantage that the signal output sounds more "present". A disadvantage is that, since the delay time is a function of input signal value, the output is no longer a linear function of the input, so non-periodic inputs can give rise to artifacts such as difference tones.

7.10 Examples

Fixed, noninterpolating delay line

Example G01.delay.pd (Figure 7.23) applies a simple delay line to an input signal. Two new objects are needed:

`delwrite~` : define and write to a delay line. The first creation argument gives the name of the delay line (and two delay lines may not share the same name). The second creation argument is the length of the delay line in milliseconds. The inlet takes an audio signal and writes it continuously into the delay line.

`delread~` : read from (or "tap") a delay line. The first creation argument gives the name of the delay line (which should agree with the name of the corresponding `delwrite~` object; this is how Pd knows which `delwrite~` to associate with the `delread~` object). The second (optional) creation argument gives the delay time in milliseconds. This may not be negative and also may not exceed the length of the delay line as specified to the `delwrite~` object. Incoming numbers (messages) may be used to change the delay time dynamically. However, this will make a discontinuous change in the output, which should therefore be muted if the delay time changes.

The example simply pairs one `delwrite~` and one `delread~` object to make a simple, noninterpolating delay. The input signal is a looped recording. The delayed and the non-delayed signal are added to make a non-recirculating comb filter. At delay times below about 10 milliseconds, the filtering effect is most prominent, and above that, a discrete echo becomes audible. There is no muting protection on the delay output, so clicks are possible when the delay time changes.

Recirculating comb filter

Example G02.delay.loop.pd (Figure 7.24) shows how to make a recirculating delay network. The delay is again accomplished with a `delwrite~`/`delread~` pair. The output of the `delread~` object is multiplied by a feedback gain of 0.7 and fed into the `delwrite~` object. An input (supplied by the `phasor~` and associated objects) is added into the `delwrite~` input; this sum becomes the output of the network. This is the recirculating comb filter of Section 7.4.

The network of tilde objects does not have any cycles, in the sense of objects feeding either directly or indirectly (via connections through other objects) to themselves. The feedback in the network occurs implicitly between the `delwrite~`and `delread~` objects.

Variable delay line

The next example, G03.delay.variable.pd (Figure 7.25), is another recirculating comb filter, this time using a variable-length delay line. One new object is introduced here:

`vd~` : Read from a delay line, with a time-varying delay time. As with the `delread~` object, this reads from a delay line whose name is specified as a creation argument. Instead of using a second argument and/or control

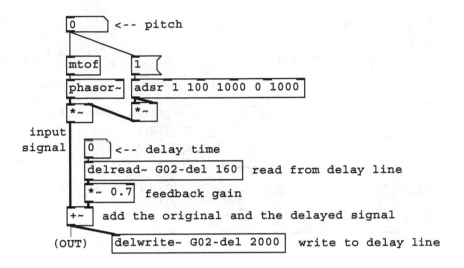

Figure 7.24: Recirculating delay (still noninterpolating).

messages to specify the delay time, for the **vd˜** object the delay in milliseconds is specified by an incoming audio signal. The delay line is read using four-point (cubic) interpolation; the minimum achievable delay is one sample.

Here the objects on the left side, from the top down to the **clip˜ -0.2 0.2** object, form a waveshaping network; the index is set by the "timbre" control, and the waveshaping output varies between a near sinusoid and a bright, buzzy sound. The output is added to the output of the **vd˜** object. The sum is then high pass filtered (the **hip˜** object at lower left), multiplied by a feedback gain, clipped, and written into the delay line at bottom right. There is a control at right to set the feedback gain; here, in contrast with the previous example, it is possible to specify a gain greater than one in order to get unstable feedback. For this reason the second **clip˜** object is inserted within the delay loop (just above the **delwrite˜** object) so that the signal cannot exceed 1 in absolute value.

The length of the delay is controlled by the signal input to the **vd˜** object. An oscillator with variable frequency and gain, in the center of the figure, provides the delay time. The oscillator is added to one to make it nonnegative before multiplying it by the "cycle depth" control, which effectively sets the range of delay times. The minimum possible delay time of 1.46 milliseconds is added so that the true range of delay times is between

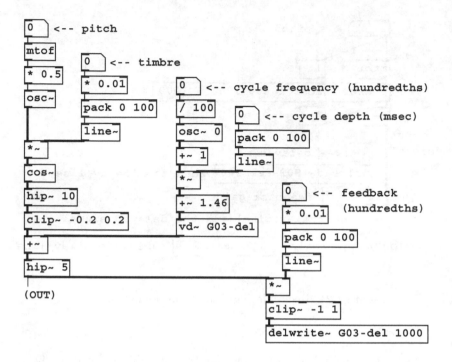

Figure 7.25: The flanger: an interpolating, variable delay line.

the minimum and the same plus twice the "depth". The reason for this
minimum delay time is taken up in the discussion of the next example.

Comb filters with variable delay times are sometimes called *flangers*. As
the delay time changes the peaks in the frequency response move up and
down in frequency, so that the timbre of the output changes constantly in
a characteristic way.

Order of execution and lower limits on delay times

When using delays (as well as other state-sharing tilde objects in Pd), the
order in which the writing and and reading operations are done can affect
the outcome of the computation. Although the tilde objects in a patch may
have a complicated topology of audio connections, in reality Pd executes
them all in a sequential order, one after the other, to compute each block
of audio output. This linear order is guaranteed to be compatible with the
audio interconnections, in the sense that no tilde object's computation is
done until all its inputs, for that same block, have been computed.

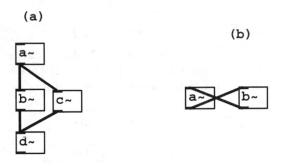

Figure 7.26: Order of execution of tilde objects in Pd: (a), an acyclic network. The objects may be executed in either the order "a-b-c-d" or "a-c-b-d". In part (b), there is a cycle, and there is thus no compatible linear ordering of the objects because each one would need to be run before the other.

Figure 7.26 shows two examples of tilde object topologies and their translation into a sequence of computation. In part (a) there are four tilde objects, and because of the connections, the object a~ must produce its output before either of b~ or c~ can run; and both of those in turn are used in the computation of d~. So the possible orderings of these four objects are "a-b-c-d" and "a-c-b-d". These two orderings will have exactly the same result unless the computation of b~ and c~ somehow affect each other's output (as delay operations might, for example).

Part (b) of the figure shows a cycle of tilde objects. This network cannot be sorted into a compatible sequential order, since each of a~ and b~ requires the other's output to be computed first. In general, a sequential ordering of the tilde objects is possible if and only if there are no cycles anywhere in the network of tilde objects and their audio signal interconnections. Pd reports an error when such a cycle appears. (Note that the situation for *control* interconnections between objects is more complicated and flexible; see the Pd documentation for details.)

To see the effect of the order of computation on a delwrite~/delread~ pair, we can write explicitly the input and output signals in the two possible orders, with the minimum possible delay. If the write operation comes first, at a block starting at sample number N, the operation can be written as:

$$x[N], \ldots, x[N + B - 1] \longrightarrow \boxed{\texttt{delwrite\textasciitilde}}$$

where B is the block size (as in Section 3.2). Having put those particular samples into the delay line, a following delread~ is able to read the same values out:

$$\boxed{\texttt{delread\~}} \longrightarrow x[N], \ldots, x[N + B - 1]$$

On the other hand, suppose the `delread~` object comes before the `delwrite~`. Then the samples $x[N], \ldots, x[N + B - 1]$ have not yet been stored in the delay line, so the most recent samples that may be read belong to the previous block:

$$\boxed{\texttt{delread\~}} \longrightarrow x[N - B], \ldots, x[N - 1]$$

$$x[N], \ldots, x[N + B - 1] \longrightarrow \boxed{\texttt{delwrite\~}}$$

Here the minimum delay we can possibly obtain is the block size B. So the minimum delay is either 0 or B, depending on the order in which the `delread~` and `delwrite~`objects are sorted into a sequence of execution.

Looking back at the patches of Figures 7.24 and 7.25, which both feature recirculating delays, the `delread~` or `vd~` object must be placed earlier in the sequence than the `delwrite~` object. This is true of any design in which a delay's output is fed back into its input. The minimum possible delay is B samples. For a (typical) sample rate of 44100 Hertz and block size of 64 samples, this comes to 1.45 milliseconds. This might not sound at first like a very important restriction. But if you are trying to tune a recirculating comb filter to a specific pitch, the highest you can get only comes to about 690 Hertz. To get shorter recirculating delays you must increase the sample rate or decrease the block size.

Example G04.control.blocksize.pd (Figure 7.27) shows how the block size can be controlled in Pd using a new object:

$\boxed{\texttt{block\~}}$, $\boxed{\texttt{switch\~}}$: Set the local block size of the patch window the object sits in. Block sizes are normally powers of two. The `switch~` object, in addition, can be used to turn audio computation within the window on and off, using control messages. Additional creation arguments can set the local sample rate and specify overlapping computations (demonstrated in Chapter 9).

In part (a) of the figure (the main patch), a rectangular pulse is sent to the `pd delay-writer` subpatch, whose output is then returned to the main patch. Part (b) shows the contents of the subpatch, which sends the pulses into a recirculating delay. The `block~` object specifies that, in this subpatch, signal computation uses a block size (B) of only one. So the minimum achievable delay is one sample instead of the default 64.

Putting a pulse (or other excitation signal) into a recirculating comb filter to make a pitch is sometimes called *Karplus-Strong synthesis*, having been described in a paper by them [KS83], although the idea seems to be older. It shows up for example in Paul Lansky's 1979 piece, *Six Fantasies on a Poem by Thomas Campion*.

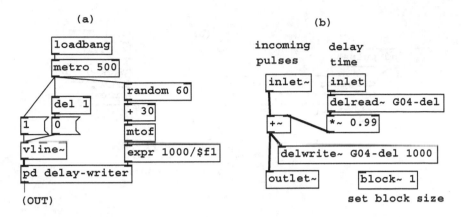

Figure 7.27: A patch using block size control to lower the loop delay below the normal 64 samples: (a) the main patch; (b) the "delay-writer" subpatch with a `block~` object and a recirculating delay network.

Order of execution in non-recirculating delay lines

In non-recirculating delay networks, it should be possible to place the operation of writing into the delay line earlier in the sequence than that of reading it. There should thus be no lower limit on the length of the delay line (except whatever is imposed by the interpolation scheme; see Section 7.7). In languages such as Csound, the sequence of unit generator computation is (mostly) explicit, so this is easy to specify. In graphical patching environments, however, the order is implicit and another approach must be taken to ensuring that, for example, a `delwrite~` object is computed before the corresponding `delread~` object. One way of accomplishing this is shown in example G05.execution.order.pd (Figure 7.28).

In part (a) of the figure, the connections in the patch do not determine which order the two delay operations appear in the sorted sequence of tilde object computation; the `delwrite~` object could be computed either before or after the `vd~` object. If we wish to make sure the writing operation happens before the reading operation, we can proceed as in part (b) of the figure and put the two operations in subpatches, connecting the two via audio signals so that the first subpatch must be computed before the second one. (Audio computation in subpatches is done atomically, in the sense that the entire subpatch contents are considered as the audio computation for the subpatch as a whole. So everything in the one subpatch happens before anything in the second one.)

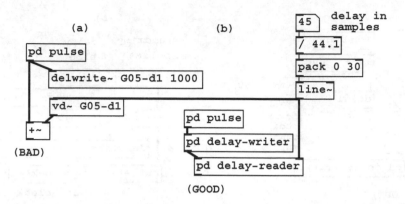

Figure 7.28: Using subpatches to ensure that delay lines are written before they are read in non-recirculating networks: (a) the `delwrite~` and `vd~` objects might be executed in either the "right" or the "wrong" order; (b) the `delwrite~` object is inside the `pd delay-writer` subpatch and the `vd~` object is inside the `pd delay-reader` one. Because of the audio connection between the two subpatches, the order of execution of the read/write pair is forced to be the correct one.

In this example, the "right" and the "wrong" way to make the comb filter have audibly different results. For delays less than 64 samples, the right hand side of the patch (using subpatches) gives the correct result, but the left hand side can't produce delays below the 64 sample block size.

Non-recirculating comb filter as octave doubler

In example G06.octave.doubler.pd (Figure 7.29) we revisit the idea of pitch-based octave shifting introduced earlier in E03.octave.divider.pd. There, knowing the periodicity of an incoming sound allowed us to tune a ring modulator to introduce subharmonics. Here we realize the octave doubler described in Section 7.3. Using a variable, non-recirculating comb filter we take out odd harmonics, leaving only the even ones, which sound an octave higher. As before, the spectral envelope of the sound is roughly preserved by the operation, so we can avoid the "chipmunk" effect we would have got by using speed change to do the transposition.

The comb filtering is done by combining two delayed copies of the incoming signal (from the `pd looper` subpatch at top). The fixed one (`delread~`) is set to the window size of the pitch following algorithm. Whereas in the earlier example this was hidden in another subpatch, we can now show

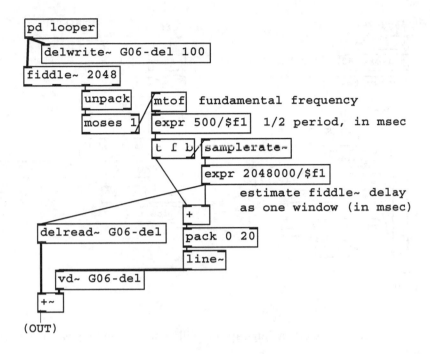

Figure 7.29: An "octave doubler" uses pitch information (obtained using a
fiddle~ object) to tune a comb filter to remove the odd harmonics in an
incoming sound.

this explicitly. The delay in milliseconds is estimated as equal to the 2048-
sample analysis window used by the fiddle~ object; in milliseconds this
comes to $1000 \cdot 2048/R$ where R is the sample rate.

   The variable delay is the same, plus 1/2 of the measured period of
the incoming sound, or $1000/(2f)$ milliseconds where f is the frequency
in cycles per second. The sum of this and the fixed delay time is then
smoothed using a line~ object to make the input signal for the variable
delay line.

   Since the difference between the two delays is $1/(2f)$, the resonant frequencies of the resulting comb filter are $2f, 4f, 6f \cdots$; the frequency response (Section 7.3) is zero at the frequencies $f, 3f, \ldots$, so the resulting
sound contains only the partials at multiples of $2f$—an octave above the
original. Seen another way, the incoming sound is output twice, a half-cycle
apart; odd harmonics are thereby shifted 180 degrees (π radians) and cancel; even harmonics are in phase with their delayed copies and remain in
the sum.

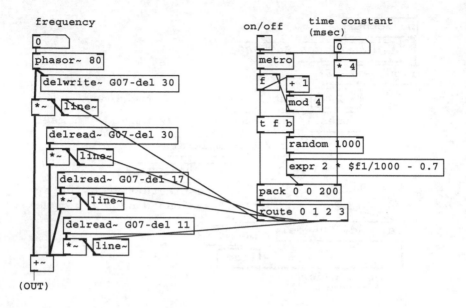

Figure 7.30: A "shaker", a four-tap comb filter with randomly varying gains on the taps.

Both this and the octave divider may be altered to make shifts of 3 or 4 to one in frequency, and they may also be combined to make compound shifts such as a musical fifth (a ratio of 3:2) by shifting down an octave and then back up a factor of three. (You should do the down-shifting before the up-shifting for best results.)

Time-varying complex comb filter: shakers

Example G07.shaker.pd (Figure 7.30) shows a different way of extending the idea of a comb filter. Here we combine the input signal at four different time shifts (instead of two, as in the original non-recirculating comb filter), each at a different positive or negative gain. To do this, we insert the input signal into a delay line and tap it at three different points; the fourth "tap" is the original, un-delayed signal.

As a way of thinking about the frequency response of a four-tap comb filter, we consider first what happens when two of the four gains are close to zero. Then we end up with a simple non-recirculating comb filter, with the slight complication that the gains of the two delayed copies may be different. If they are both of the same sign, we get the same peaks and valleys as predicted in Section 7.3, only with the valleys between peaks possibly more shallow. If they are opposite in sign, the valleys become peaks and the peaks become valleys.

Depending on which two taps we supposed were nonzero, the peaks and valleys are spaced by different amounts; the delay times are chosen so that 6 different delay times can arise in this way, ranging between 6 and 30 milliseconds. In the general case in which all the gains are non-zero, we can imagine the frequency response function varying continuously between these extremes, giving a succession of complicated patterns. This has the effect of raising and lowering the amplitudes of the partials of the incoming signal, all independently of the others, in a complicated pattern, to give a steadily time-varying timbre.

The right-hand side of the patch takes care of changing the gains of the input signal and its three time-shifted copies. Each time the `metro` object fires, a counter is incremented (the `f`, `+ 1`, and `mod 4` objects). This controls which of the amplitudes will be changed. The amplitude itself is computed by making a random number and normalizing it to lie between -0.7 and 1.3 in value. The random value and the index are packed (along with a third value, a time interval) and this triple goes to the `route` object. The first element of the triple (the counter) selects which output to send the other two values to; as a result, one of the four possible `line~` objects gets a message to ramp to a new value.

If the time variation is done quickly enough, there is also a modulation effect on the original signal; in this situation the straight line segments used in this example should be replaced by modulating signals with more controllable frequency content, for instance using filters (the subject of Chapter 8).

Reverberator

Example G08.reverb.pd (Figure 7.31) shows a simple artificial reverberator, essentially a realization of the design shown in Figure 7.15. Four delay lines are fed by the input and by their own recirculated output. The delay outputs are intermixed using rotation matrices, built up from elementary rotations of $\pi/4$ as in Figure 7.13 (part a).

The normalizing multiplication (by $\sqrt{1/2}$ at each stage) is absorbed into the feedback gain, which therefore cannot exceed 1/2. At a feedback gain of exactly 1/2, all the energy leaving the delay lines is reinserted into them, so the reverberation lasts perpetually.

Figure 7.32 shows the interior of the `reverb-echo` abstraction used in the reverberator. The two inputs are mixed (using the same rotation matrix and again leaving the renormalization for later). One channel is

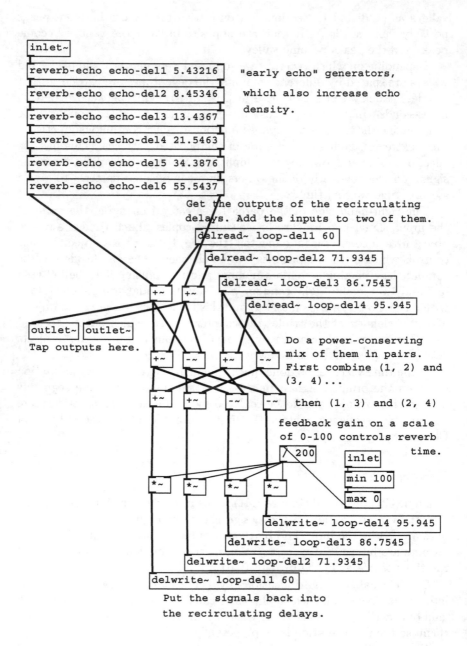

Figure 7.31: An artificial reverberator.

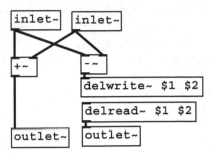

Figure 7.32: The echo generator used in the reverberator.

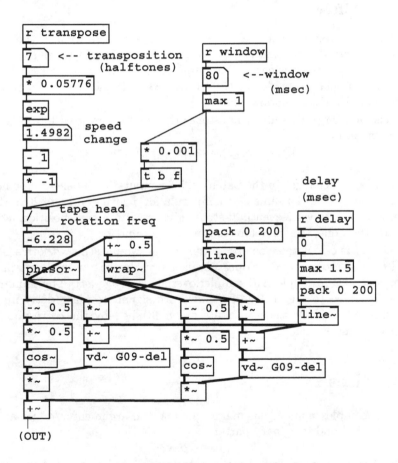

Figure 7.33: A pitch shifter using two variable taps into a delay line.

then delayed. The delay times are selected to grow roughly exponentially; this ensures a smooth and spread-out pattern of echos.

Many extensions of this idea are possible of which we'll only name a few. It is natural, first, to put low-pass filters at the end of the delay lines, to mimic the typically faster decay of high frequencies than low ones. It is also common to use more than four recirculating delays; one reverberator in the Pd distribution uses sixteen. Finally, it is common to allow separate control of the amplitudes of the early echos (heard directly) and that of the recirculating signal; parameters such as these are thought to control sonic qualities described as "presence", "warmth", "clarity", and so on.

Pitch shifter

Example G09.pitchshift.pd (Figure 7.33) shows a realization of the pitch shifter described in Section 7.9. A delay line (defined and written elsewhere in the patch) is read using two vd~ objects. The delay times vary between a minimum delay (provided as the "delay" control) and the minimum plus a window size (the "window" control.)

The desired pitch shift in half-tones (h) is first converted into a transposition factor

$$t = 2^{h/12} = e^{\log(2)/12 \cdot h} \approx e^{0.05776h}$$

(called "speed change" in the patch). The computation labeled "tape head rotation speed" is the same as the formula for f given on Page 208. Here the positive interval (seven half-steps) gives rise to a transposition factor greater than one, and therefore to a negative value for f.

Once f is calculated, the production of the two phased sawtooth signals and the corresponding envelopes parallels exactly that of the overlapping sample looper (example B10.sampler.overlap.pd, Page 54). The minimum delay is added to each of the two sawtooth signals to make delay inputs for the vd~ objects, whose outputs are multiplied by the corresponding envelopes and summed.

Exercises

1. A complex number has magnitude one and argument $\pi/4$. What are its real and imaginary parts?

2. A complex number has magnitude one and real part $1/2$. What is its imaginary part? (There are two possible values.)

3. What delay time would you give a comb filter so that its first frequency response peak is at 440 Hertz? If the sample rate is 44100, what frequency would correspond to the nearest integer delay?

4. Suppose you made a variation on the non-recirculating comb filter so that the delayed signal was subtracted from the original instead of adding. What would the new frequency response be?

5. If you want to make a 6-Hertz vibrato with a sinusoidally varying delay line, and if you want the vibrato to change the frequency by 5%, how big a delay variation would you need? How would this change if the same depth of vibrato was desired at 12 Hertz?

6. A complex sinusoid $X[n]$ has frequency 11025 Hertz, amplitude 50 and initial phase 135 degrees. Another one, $Y[n]$, has the same frequency, but amplitude 20 and initial phase 45 degrees. What are the amplitude and initial phase of the sum of X and Y?

7. What are the frequency, initial phase, and amplitude of the signal obtained when $X[n]$ (above) is delayed 4 samples?

8. Show that the frequency response of a recirculating comb filter with delay time d and feedback gain g, as a function of angular frequency ω, is equal to:

$$\left[(1 - g\cos(\omega d))^2 + (g\sin(\omega d))^2 \right]^{-1/2}$$

Chapter 8

Filters

In the previous chapter we saw that a delay network can have a non-uniform frequency response—a gain that varies as a function of frequency. Delay networks also typically change the phase of incoming signals variably depending on frequency. When the delay times used are very short, the most important properties of a delay network become its frequency and phase response. A delay network that is designed specifically for its frequency or phase response is called a *filter*.

In block diagrams, filters are shown as in Figure 8.1 (part a). The curve shown within the block gives a qualitative representation of the filter's frequency response. The frequency response may vary with time, and depending on the design of the filter, one or more controls (or additional audio inputs) might be used to change it.

Suppose, following the procedure of Section 7.3, we put a unit-amplitude, complex-valued sinusoid with angular frequency ω into a filter. We expect to get out a sinusoid of the same frequency and some amplitude, which depends on ω. This gives us a complex-valued function $H(\omega)$, which is called the *transfer function* of the filter.

The frequency response is the gain as a function of the frequency ω. It is is equal to the magnitude of the transfer function. A filter's frequency response is customarily graphed as in Figure 8.1 (part b). An incoming unit-amplitude sinusoid of frequency ω comes out of the filter with magnitude $|H(\omega)|$.

It is sometimes also useful to know the phase response of the filter, equal to $\angle(H(\omega))$. For a fixed frequency ω, the filter's output phase will be $\angle(H(\omega))$ radians ahead of its input phase.

The design and use of filters is a huge subject, because the wide range of uses a filter might be put to suggests a wide variety of filter design processes.

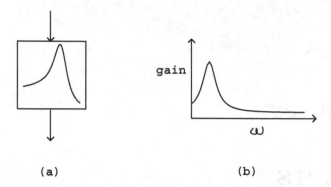

Figure 8.1: Representations of a filter: (a) in a block diagram; (b) a graph of its frequency response.

In some applications a filter must exactly follow a prescribed frequency response, in others it is important to minimize computation time, in others the phase response is important, and in still others the filter must behave well when its parameters change quickly with time.

8.1   Taxonomy of Filters

Over the history of electronic music the technology for building filters has changed constantly, but certain kinds of filters reappear often. In this section we will give some nomenclature for describing filters of several generic, recurring types. Later we'll develop some basic strategies for making filters with desired characteristics, and finally we'll discuss some common applications of filters in electronic music.

8.1.1   Low-pass and high-pass filters

By far the most frequent purpose for using a filter is extracting either the low-frequency or the high-frequency portion of an audio signal, attenuating the rest. This is accomplished using a *low-pass* or *high-pass* filter.

Ideally, a low-pass or high-pass filter would have a frequency response of one up to (or down to) a specified cutoff frequency and zero past it; but such filters cannot be realized in practice. Instead, we try to find realizable approximations to this ideal response. The more design effort and computation time we put into it, the closer we can get.

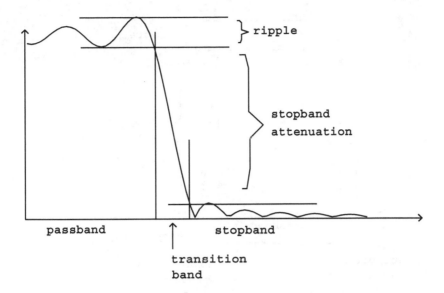

Figure 8.2: Terminology for describing the frequency response of low-pass and high-pass filters. The horizontal axis is frequency and the vertical axis is gain. A low-pass filter is shown; a high-pass filter has the same features switched from right to left.

Figure 8.2 shows the frequency response of a low-pass filter. Frequency is divided into three bands, labeled on the horizontal axis. The *passband* is the region (frequency band) where the filter should pass its input through to its output with unit gain. For a low-pass filter (as shown), the passband reaches from a frequency of zero up to a certain frequency limit. For a high-pass filter, the passband would appear on the right-hand side of the graph and would extend from the frequency limit up to the highest frequency possible. Any realizable filter's passband will be only approximately flat; the deviation from flatness is called the *ripple*, and is often specified by giving the ratio between the highest and lowest gain in the passband, expressed in decibels. The ideal low-pass or high-pass filter would have a ripple of 0 dB.

The *stopband* of a low-pass or high-pass filter is the frequency band over which the filter is intended not to transmit its input. The *stopband attenuation* is the difference, in decibels, between the lowest gain in the passband and the highest gain in the stopband. Ideally this would be infinite; the higher the better.

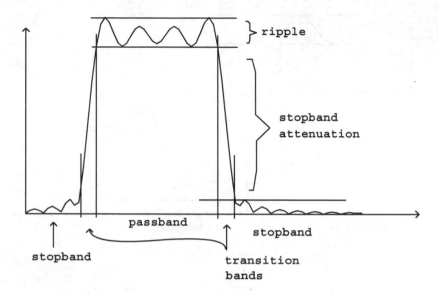

Figure 8.3: Terminology for describing the frequency response of band-pass and stop-band filters. The horizontal axis is frequency and the vertical axis is gain. A band-pass filter is shown; a stop-band filter would have a contiguous stopband surrounded by two passbands.

Finally, a realizable filter, whose frequency response is always a continuous function of frequency, must have a frequency band over which the gain drops from the passband gain to the stopband gain; this is called the *transition band.* The thinner this band can be made, the more nearly ideal the filter.

8.1.2   Band-pass and stop-band filters

A *band-pass filter* admits frequencies within a given band, rejecting frequencies below it and above it. Figure 8.3 shows the frequency response of a band-pass filter, with the key parameters labelled. A stop-band filter does the reverse, rejecting frequencies within the band and letting through frequencies outside it.

In practice, a simpler language is often used for describing bandpass filters, as shown in Figure 8.4. Here there are only two parameters: a *center frequency* and a *bandwidth.* The passband is considered to be the region where the filter has at least half the power gain as at the peak (i.e.,

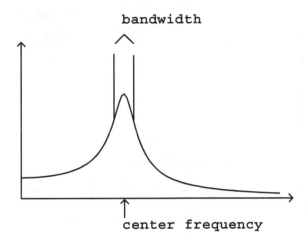

Figure 8.4: A simplified view of a band-pass filter, showing bandwidth and center frequency.

the gain is within 3 decibels of its maximum). The bandwidth is the width, in frequency units, of the passband. The center frequency is the point of maximum gain, which is approximately the midpoint of the passband.

8.1.3 Equalizing filters

In some applications, such as *equalization*, the goal isn't to pass signals of certain frequencies while stopping others altogether, but to make controllable adjustments, boosting or attenuating a signal, over a frequency range, by a desired gain. Two filter types are useful for this. First, a *shelving filter* (Figure 8.5) is used for selectively boosting or reducing either the low or high end of the frequency range. Below a selectable crossover frequency, the filter tends toward a low-frequency gain, and above it it tends toward a different, high-frequency one. The crossover frequency, low-frequency gain, and high-frequency gain can all be adjusted independently.

Second, a *peaking filter* (Figure 8.6) is capable of boosting or attenuating signals within a range of frequencies. The center frequency and bandwidth (which together control the range of frequencies affected), and the in-band and out-of-band gains are separately adjustible.

Parametric equalizers often employ two shelving filters (one each to adjust the low and high ends of the spectrum) and two or three peaking filters to adjust bands in between.

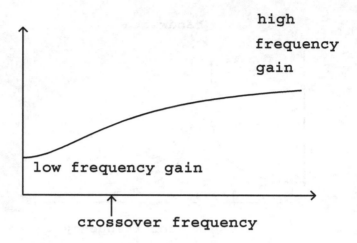

Figure 8.5: A shelving filter, showing low and high frequency gain, and crossover frequency.

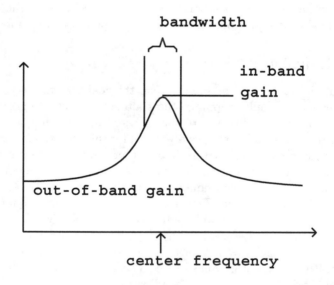

Figure 8.6: A peaking filter, with controllable center frequency, bandwidth, and in-band and out-of-band gains.

8.2 Elementary Filters

We saw in Chapter 7 how to predict the frequency and phase response of delay networks. The art of filter design lies in finding a delay network whose transfer function (which controls the frequency and phase response) has a desired shape. We will develop an approach to building such delay networks out of the two types of comb filters developed in Chapter 7: recirculating and non-recirculating. Here we will be interested in the special case where the delay is only one sample in length. In this situation, the frequency responses shown in Figures 7.6 and 7.10 no longer look like combs; the second peak recedes all the way to the sample rate, 2π radians, when $d = 1$. Since only frequencies between 0 and the Nyquist frequency (π radians) are audible, in effect there is only one peak when $d = 1$.

In the comb filters shown in Chapter 7, the peaks are situated at DC (zero frequency), but we will often wish to place them at other, nonzero frequencies. This is done using delay networks—comb filters—with complex-valued gains.

8.2.1 Elementary non-recirculating filter

The non-recirculating comb filter may be generalized to yield the design shown in Figure 8.7. This is the *elementary non-recirculating filter*, of the

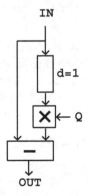

Figure 8.7: A delay network with a single-sample delay and a complex gain Q. This is the non-recirculating elementary filter, first form. Compare the non-recirculating comb filter shown in Figure 7.3, which corresponds to choosing $Q = -1$ here.

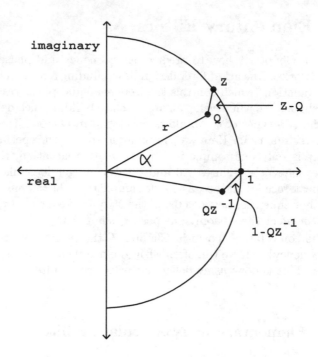

Figure 8.8: Diagram for calculating the frequency response of the non-recirculating elementary filter (Figure 8.7). The frequency response is given by the length of the segment connecting Z to Q in the complex plane.

first form. Its single, complex-valued parameter Q controls the complex gain of the delayed signal subtracted from the original one.

To find its frequency response, as in Chapter 7 we feed the delay network a complex sinusoid $1, Z, Z^2, \ldots$ whose frequency is $\omega = \arg(Z)$. The nth sample of the input is Z^n and that of the output is

$$(1 - QZ^{-1})Z^n$$

so the transfer function is

$$H(Z) = 1 - QZ^{-1}$$

This can be analyzed graphically as shown in Figure 8.8. The real numbers r and α are the magnitude and argument of the complex number Q:

$$Q = r \cdot (\cos(\alpha) + i\sin(\alpha))$$

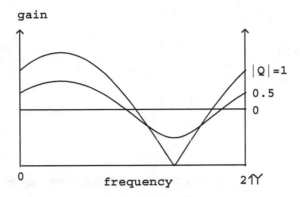

Figure 8.9: Frequency response of the elementary non-recirculating filter Figure 8.7. Three values of Q are used, all with the same argument (-2 radians), but with varying absolute value (magnitude) $r = |Q|$.

The gain of the filter is the distance from the point Q to the point Z in the complex plane. Analytically we can see this because

$$|1 - QZ^{-1}| = |Z||1 - QZ^{-1}| = |Q - Z|$$

Graphically, the number QZ^{-1} is just the number Q rotated backwards (clockwise) by the angular frequency ω of the incoming sinusoid. The value $|1 - QZ^{-1}|$ is the distance from QZ^{-1} to 1 in the complex plane, which is equal to the distance from Q to Z.

As the frequency of the input sweeps from 0 to 2π, the point Z travels couterclockwise around the unit circle. At the point where $\omega = \alpha$, the distance is at a minimum, equal to $1 - r$. The maximum occurs which Z is at the opposite point of the circle. Figure 8.9 shows the transfer function for three different values of $r = |Q|$.

8.2.2 Non-recirculating filter, second form

Sometimes we will need a variant of the filter above, shown in Figure 8.10, called the *elementary non-recirculating filter, second form*. Instead of multiplying the delay output by Q we multiply the direct signal by its *complex conjugate* $\overline{Q}$. If

$$A = a + bi = r \cdot (\cos(\alpha) + i\sin(\alpha))$$

is any complex number, its complex conjugate is defined as:

$$\overline{A} = a - bi = r \cdot (\cos(\alpha) - i\sin(\alpha))$$

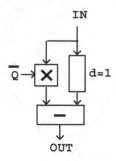

Figure 8.10: The elementary non-recirculating filter, second form.

Graphically this reflects points of the complex plane up and down across the real axis. The transfer function of the new filter is

$$H(Z) = \overline{Q} - Z^{-1}$$

This gives rise to the same frequency response as before since

$$|\overline{Q} - Z^{-1}| = |Q - \overline{Z^{-1}}| = |Q - Z|$$

Here we use the fact that $\overline{Z} = Z^{-1}$, for any unit complex number Z, as can be verified by writing out $Z\overline{Z}$ in either polar or rectangular form.

Although the two forms of the elementary non-recirculating filter have the same frequency response, their phase responses are different; this will occasionally lead us to prefer the second form.

8.2.3   Elementary recirculating filter

The *elementary recirculating filter* is the recirculating comb filter of Figure 7.7 with a complex-valued feedback gain P as shown in Figure 8.11 (part a). By the same analysis as before, feeding this network a sinusoid whose nth sample is Z^n gives an output of:

$$\frac{1}{1 - PZ^{-1}} Z^n$$

so the transfer function is

$$H(Z) = \frac{1}{1 - PZ^{-1}}$$

The recirculating filter is stable when $|P| < 1$; when, instead, $|P| > 1$ the output grows exponentially as the delayed sample recirculates.

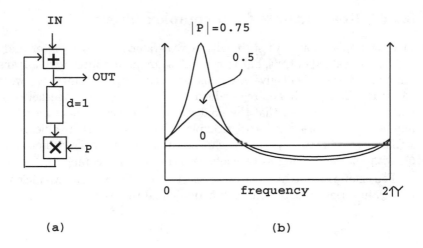

Figure 8.11: The elementary recirculating filter: (a) block diagram; (b) frequency response.

The transfer function is thus just the inverse of that of the non-recirculating filter (first form). If you put the two in series with $P = Q$, the output theoretically equals the input. (This analysis only demonstrates it for sinusoidal inputs; that it follows for other signals as well can be verified by working out the impulse response of the combined network).

8.2.4   Compound filters

We can use the recirculating and non-recirculating filters developed here to create a *compound filter* by putting several elementary ones in series. If the parameters of the non-recirculating ones (of the first type) are $Q_1, \ldots, Q_j$ and those of the recirculating ones are $P_1, \ldots, P_k$, then putting them all in series, in any order, will give the transfer function:

$$H(Z) = \frac{(1 - Q_1 Z^{-1}) \cdots (1 - Q_j Z^{-1})}{(1 - P_1 Z^{-1}) \cdots (1 - P_k Z^{-1})}$$

The frequency response of the resulting compound filter is the product of those of the elementary ones. (One could also combine elementary filters by adding their outputs, or making more complicated networks of them; but for most purposes the series configuration is the easiest one to work with.)

8.2.5   Real outputs from complex filters

In most applications, we start with a real-valued signal to filter and we need a real-valued output, but in general, a compound filter with a transfer function as above will give a complex-valued output. However, we can construct filters with non-real-valued coefficients which nonetheless give real-valued outputs, so that the analysis that we carry out using complex numbers can be used to predict, explain, and control real-valued output signals. We do this by pairing each elementary filter (with coefficient P or Q) with another having as its coefficient the complex conjugate $\overline{P}$ or $\overline{Q}$.

For example, putting two non-recirculating filters, with coefficients Q and $\overline{Q}$, in series gives a transfer function equal to:

$$H(Z) = (1 - QZ^{-1}) \cdot (1 - \overline{Q}Z^{-1})$$

which has the property that:

$$H(\overline{Z}) = \overline{H(Z)}$$

Now if we put any real-valued sinusoid:

$$X_n = 2\operatorname{re}(AZ^n) = AZ^n + \overline{AZ}^n$$

we get out:

$$A \cdot H(Z) \cdot Z^n + \overline{A} \cdot \overline{H(Z)} \cdot \overline{Z}^n$$

which, by inspection, is another real sinusoid. Here we're using two properties of complex conjugates. First, you can add and multiply them at will:

$$\overline{A + B} = \overline{A} + \overline{B}$$

$$\overline{AB} = \overline{A} \cdot \overline{B}$$

and second, anything plus its complex conjugate is real, and is in fact twice its real part:

$$A + \overline{A} = 2\operatorname{re}(A)$$

This result for two conjugate filters extends to any compound filter; in general, we always get a real-valued output from a real-valued input if we arrange that each coefficient Q_i and P_i in the compound filter is either real-valued, or else appears in a pair with its complex conjugate.

8.2.6 Two recirculating filters for the price of one

When pairing recirculating elementary filters, it is possible to avoid computing one of each pair, as long as the input is real-valued (and so, the output is as well.) Supposing the input is a real sinusoid of the form:

$$AZ^n + \overline{A}Z^{-n}$$

and we apply a single recirculating filter with coefficient P. Letting $a[n]$ denote the real part of the output, we have:

$$a[n] = \mathrm{re}\left[\frac{1}{1 - PZ^{-1}}AZ^n + \frac{1}{1 - PZ}\overline{A}Z^{-n}\right]$$

$$= \mathrm{re}\left[\frac{1}{1 - PZ^{-1}}AZ^n + \frac{1}{1 - \overline{P}Z^{-1}}AZ^n\right]$$

$$= \mathrm{re}\left[\frac{2 - 2\,\mathrm{re}(P)Z^{-1}}{(1 - PZ^{-1})(1 - \overline{P}Z^{-1})}AZ^n\right]$$

$$= \mathrm{re}\left[\frac{1 - \mathrm{re}(P)Z^{-1}}{(1 - PZ^{-1})(1 - \overline{P}Z^{-1})}AZ^n + \frac{1 - \mathrm{re}(P)\overline{Z}^{-1}}{(1 - \overline{PZ}^{-1})(1 - P\overline{Z}^{-1})}\overline{A}Z^{-n}\right]$$

(In the second step we used the fact that you can conjugate all or part of an expression, without changing the result, if you're just going to take the real part anyway. The fourth step did the same thing backward.) Comparing the input to the output, we see that the effect of passing a real signal through a complex one-pole filter, then taking the real part, is equivalent to passing the signal through a two-pole, one-zero filter with transfer function equal to:

$$H_{\mathrm{re}}(Z) = \frac{1 - \mathrm{re}(P)Z^{-1}}{(1 - PZ^{-1})(1 - \overline{P}Z^{-1})}$$

A similar calculation shows that taking the imaginary part (considered as a real signal) is equivalent to filtering the input with the transfer function:

$$H_{\mathrm{im}}(Z) = \frac{\mathrm{im}(P)Z^{-1}}{(1 - PZ^{-1})(1 - \overline{P}Z^{-1})}$$

So taking either the real or imaginary part of a one-pole filter output gives filters with two conjugate poles. The two parts can be combined to synthesize filters with other possible numerators; in other words, with one complex recirculating filter we can synthesize a filter that acts on real signals with two (complex conjugate) poles and one (real) zero.

This technique, known as *partial fractions*, may be repeated for any number of stages in series as long as we compute the appropriate combination of real and imaginary parts of the output of each stage to form the (real) input of the next stage. No similar shortcut seems to exist for non-recirculating filters; for them it is necessary to compute each member of each complex-conjugate pair explicitly.

8.3   Designing Filters

The frequency response of a series of elementary recirculating and non-recirculating filters can be estimated graphically by plotting all the coefficients $Q_1, \ldots, Q_j$ and $P_1, \ldots, P_k$ on the complex plane and reasoning as in Figure 8.8. The overall frequency response is the product of all the distances from the point Z to each of the Q_i, divided by the product of the distances to each of the P_i.

One customarily marks each of the Q_i with an "o" (calling it a "zero") and each of the P_i with an "x" (a "pole"); their names are borrowed from the field of complex analysis. A plot showing the poles and zeroes associated with a filter is unimaginatively called a *pole-zero plot*.

When Z is close to a zero the frequency response tends to dip, and when it is close to a pole, the frequency response tends to rise. The effect of a pole or a zero is more pronounced, and also more local, if it is close to the unit circle that Z is constrained to lie on. Poles must lie within the unit circle for a stable filter. Zeros may lie on or outside it, but any zero Q outside the unit circle may be replaced by one within it, at the point $1/\overline{Q}$, to give a constant multiple of the same frequency response. Except in special cases we will keep the zeros inside the circle as well as the poles.

In the rest of this section we will show how to construct several of the filter types most widely used in electronic music. The theory of digital filter design is vast, and we will only give an introduction here. A deeper treatment is available online from Julius Smith at ccrma.stanford.edu. See also [Ste96] for an introduction to filter design from the more general viewpoint of digital signal processing.

8.3.1   One-pole low-pass filter

The one-pole low-pass filter has a single pole located at a positive real number p, as pictured in Figure 8.12. This is just a recirculating comb filter with delay length $d = 1$, and the analysis of Section 7.4 applies. The maximum gain occurs at a frequency of zero, corresponding to the point on the circle closest to the point p. The gain there is $1/(1-p)$. Assuming p is

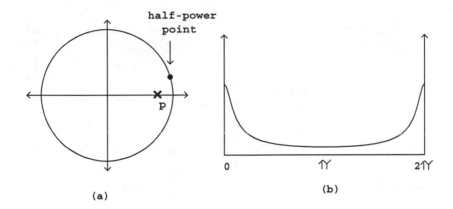

Figure 8.12: One-pole low-pass filter: (a) pole-zero diagram; (b) frequency response.

close to one, if we move a distance of $1 - p$ units up or down from the real (horizontal) axis, the distance increases by a factor of about $\sqrt{2}$, and so we expect the half-power point to occur at an angular frequency of about $1 - p$.

This calculation is often made in reverse: if we wish the half-power point to lie at a given angular frequency ω, we set $p = 1 - \omega$. This approximation only works well if the value of ω is well under $\pi/2$, as it often is in practice. It is customary to normalize the one-pole low-pass filter, multiplying it by the constant factor $1 - p$ in order to give a gain of 1 at zero frequency; nonzero frequencies will then get a gain less than one.

The frequency response is graphed in Figure 8.12 (part b). The audible frequencies only reach to the middle of the graph; the right-hand side of the frequency response curve all lies above the Nyquist frequency π.

The one-pole low-pass filter is often used to seek trends in noisy signals. For instance, if you use a physical controller and only care about changes on the order of $1/10$ second or so, you can smooth the values with a low-pass filter whose half-power point is 20 or 30 cycles per second.

8.3.2 One-pole, one-zero high-pass filter

Sometimes an audio signal carries an unwanted constant offset, or in other words, a zero-frequency component. For example, the waveshaping spectra of Section 5.3 almost always contain a constant component. This is inaudible, but, since it specifies electrical power that is sent to your speakers,

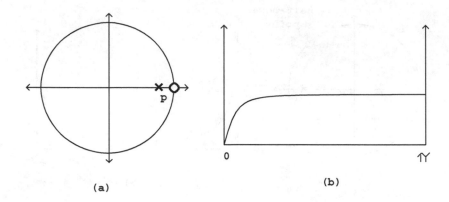

Figure 8.13: One-pole, one-zero high-pass filter: (a) pole-zero diagram;
(b) frequency response (from zero to Nyquist frequency).

its presence reduces the level of loudness you can reach without distortion.
Another name for a constant signal component is "DC", meaning "direct
current".

An easy and practical way to remove the zero-frequency component
from an audio signal is to use a one-pole low-pass filter to extract it, and
then subtract the result from the signal. The resulting transfer function is
one minus the transfer function of the low-pass filter:

$$H(Z) = 1 - \frac{1-p}{1-pZ^{-1}} = p\frac{1-Z^{-1}}{1-pZ^{-1}}$$

The factor of $1-p$ in the numerator of the low-pass transfer function is the
normalization factor needed so that the gain is one at zero frequency.

By examining the right-hand side of the equation (comparing it to the
general formula for compound filters), we see that there is still a pole at
the real number p, and there is now also a zero at the point 1. The pole-
zero plot is shown in Figure 8.13 (part a), and the frequency response in
part (b). (Henceforth, we will only plot frequency responses to the Nyquist
frequency π; in the previous example we plotted it all the way up to the
sample rate, 2π.)

8.3.3   Shelving filter

Generalizing the one-zero, one-pole filter above, suppose we place the zero
at a point q, a real number close to, but less than, one. The pole, at the

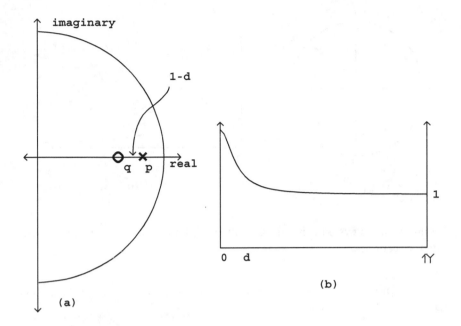

Figure 8.14: One-pole, one-zero shelving filter: (a) pole-zero diagram; (b) frequency response.

point p, is similarly situated, and might be either greater than or less than q, i.e., to the right or left, respectively, but with both q and p within the unit circle. This situation is diagrammed in Figure 8.14.

At points of the circle far from p and q, the effects of the pole and the zero are nearly inverse (the distances to them are nearly equal), so the filter passes those frequencies nearly unaltered. In the neighborhood of p and q, on the other hand, the filter will have a gain greater or less than one depending on which of p or q is closer to the circle. This configuration therefore acts as a low-frequency shelving filter. (To make a high-frequency shelving filter we do the same thing, only placing p and q close to -1 instead of 1.)

To find the parameters of a shelving filter given a desired transition frequency ω (in angular units) and low-frequency gain g, first we choose an average distance d, as pictured in the figure, from the pole and the zero to the edge of the circle. For small values of d, the region of influence is about d radians, so simply set $d = \omega$ to get the desired transition frequency.

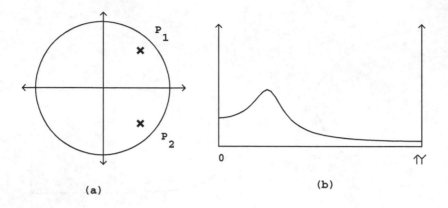

Figure 8.15: Two-pole band-pass filter: (a) pole-zero diagram; (b) frequency response.

Then put the pole at $p = 1 - d/\sqrt{g}$ and the zero at $q = 1 - d\sqrt{g}$. The gain at zero frequency is then

$$\frac{1-q}{1-p} = g$$

as desired. For example, in the figure, d is 0.25 radians and g is 2.

8.3.4   Band-pass filter

Starting with the three filter types shown above, which all have real-valued poles and zeros, we now transform them to operate on bands located off the real axis. The low-pass, high-pass, and shelving filters will then become band-pass, stop-band, and peaking filters. First we develop the band-pass filter. Suppose we want a center frequency at ω radians and a bandwidth of β. We take the low-pass filter with cutoff frequency β; its pole is located, for small values of β, roughly at $p = 1 - \beta$. Now rotate this value by ω radians in the complex plane, i.e., multiply by the complex number $\cos\omega + i\sin\omega$. The new pole is at:

$$P_1 = (1 - \beta)(\cos\omega + i\sin\omega)$$

To get a real-valued output, this must be paired with another pole:

$$P_2 = \overline{P_1} = (1 - \beta)(\cos\omega - i\sin\omega)$$

The resulting pole-zero plot is as shown in Figure 8.15.

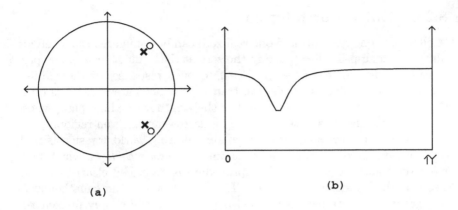

Figure 8.16: A peaking filter: (a) pole-zero diagram; (b) frequency response. Here the filter is set to attenuate by 6 decibels at the center frequency.

The peak is approximately (not exactly) at the desired center frequency ω, and the frequency response drops by 3 decibels approximately β radians above and below it. It is often desirable to normalize the filter to have a peak gain near unity; this is done by multiplying the input or output by the product of the distances of the two poles to the peak on the circle, or (very approximately):

$$\beta * (\beta + 2\omega)$$

For some applications it is desirable to add a zero at the points 1 and -1, so that the gain drops to zero at angular frequencies 0 and π.

8.3.5 Peaking and stop-band filter

In the same way, a peaking filter is obtained from a shelving filter by rotating the pole and the zero, and by providing a conjugate pole and zero, as shown in Figure 8.16. If the desired center frequency is ω, and the radii of the pole and zero (as for the shelving filter) are p and q, then we place the the upper pole and zero at

$$P_1 = p \cdot (\cos\omega + i\sin\omega)$$

$$Q_1 = q \cdot (\cos\omega + i\sin\omega)$$

As a special case, placing the zero on the unit circle gives a stop-band filter; in this case the gain at the center frequency is zero. This is analogous to the one-pole, one-zero high-pass filter above.

8.3.6   Butterworth filters

A filter with one real pole and one real zero can be configured as a shelving filter, as a high-pass filter (putting the zero at the point 1) or as a low-pass filter (putting the zero at -1). The frequency responses of these filters are quite blunt; in other words, the transition regions are wide. It is often desirable to get a sharper filter, either shelving, low- or high-pass, whose two bands are flatter and separated by a narrower transition region.

A procedure borrowed from the analog filtering world transforms real, one-pole, one-zero filters to corresponding *Butterworth filters*, which have narrower transition regions. This procedure is described clearly and elegantly in the last chapter of [Ste96]. The derivation uses more mathematics background than we have developed here, and we will simply present the result without deriving it.

To make a Butterworth filter out of a high-pass, low-pass, or shelving filter, suppose that either the pole or the zero is given by the expression

$$\frac{1 - r^2}{(1 + r)^2}$$

where r is a parameter ranging from 1 to ∞. If $r = 0$ this is the point 1, and if $r = \infty$ it's -1.

Then, for reasons which will remain mysterious, we replace the point (whether pole or zero) by n points given by:

$$\frac{(1 - r^2) - (2r\sin(\alpha))i}{1 + r^2 + 2r\cos(\alpha))}$$

where α ranges over the values:

$$\frac{\pi}{2}(\frac{1}{n} - 1), \ \frac{\pi}{2}(\frac{3}{n} - 1), \ \ldots, \ \frac{\pi}{2}(\frac{2n - 1}{n} - 1)$$

In other words, α takes on n equally spaced angles between $-\pi/2$ and $\pi/2$. The points are arranged in the complex plane as shown in Figure 8.17. They lie on a circle through the original real-valued point, which cuts the unit circle at right angles.

A good estimate for the cutoff or transition frequency defined by these circular collections of poles or zeros is simply the spot where the circle intersects the unit circle, corresponding to $\alpha = \pi/2$. This gives the point

$$\frac{(1 - r^2) - 2ri}{1 + r^2}$$

which, after some algebra, gives an angular frequency equal to

$$\beta = 2\arctan(r)$$

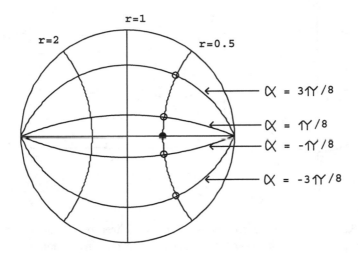

Figure 8.17: Replacing a real-valued pole or zero (shown as a solid dot) with an array of four of them (circles) as for a Butterworth filter. In this example we get four new poles or zeros as shown, lying along the circle where $r = 0.5$.

Figure 8.18 (part a) shows a pole-zero diagram and frequency response for a Butterworth low-pass filter with three poles and three zeros. Part (b) shows the frequency response of the low-pass filter and three other filters obtained by choosing different values of β (and hence r) for the zeros, while leaving the poles stationary. As the zeros progress from $\beta = \pi$ to $\beta = 0$, the filter, which starts as a low-pass filter, becomes a shelving filter and then a high-pass one.

8.3.7 Stretching the unit circle with rational functions

In Section 8.3.4 we saw a simple way to turn a low-pass filter into a band-pass one. It is tempting to apply the same method to turn our Butterworth low-pass filter into a higher-quality band-pass filter; but if we wish to preserve the high quality of the Butterworth filter we must be more careful than before in the design of the transformation used. In this section we will prepare the way to making the Butterworth band-pass filter by introducing a class of rational transformations of the complex plane which preserve the unit circle.

This discussion is adapted from [PB87], pp. 201-206 (I'm grateful to Julius Smith for this pointer). There the transformation is carried out

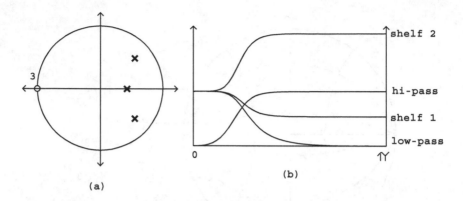

Figure 8.18: Butterworth low-pass filter with three poles and three zeros:
(a) pole-zero plot. The poles are chosen for a cutoff frequency $\beta = \pi/4$;
(b) frequency responses for four filters with the same pole configuration,
with different placements of zeros (but leaving the poles fixed). The low-
pass filter results from setting $\beta = \pi$ for the zeros; the two shelving filters
correspond to $\beta = 3\pi/10$ and $\beta = 2\pi/10$, and finally the high-pass filter is
obtained setting $\beta = 0$. The high-pass filter is normalized for unit gain at
the Nyquist frequency, and the others for unit gain at DC.

in continuous time, but here we have adapted the method to operate in
discrete time, in order to make the discussion self-contained.

The idea is to start with any filter with a transfer function as before:

$$H(Z) = \frac{(1 - Q_1 Z^{-1}) \cdots (1 - Q_j Z^{-1})}{(1 - P_1 Z^{-1}) \cdots (1 - P_k Z^{-1})}$$

whose frequency response (the gain at a frequency ω) is given by:

$$|H(\cos(\omega) + i \sin(\omega))|$$

Now suppose we can find a rational function, $R(Z)$, which distorts the
unit circle in some desirable way. For R to be a rational function means
that it can be written as a quotient of two polynomials (for example, the
transfer function H is a rational function). That R sends points on the unit
circle to other points on the unit circle is just the condition that $|R(Z)| = 1$
whenever $Z = 1$. It can easily be checked that any function of the form

$$R(Z) = U \cdot \frac{A_n Z^n + A_{n-1} Z^{n-1} + \cdots + A_0}{\overline{A_0} Z^n + \overline{A_1} Z^{n-1} + \cdots + \overline{A_n}}$$

(where $|U| = 1$) has this property. The same reasoning as in Section 8.2.2 confirms that $|R(Z)| = 1$ whenever $Z = 1$.

Once we have a suitable rational function R, we can simply compose it with the original transfer function H to fabricate a new rational function,

$$J(Z) = H(R(Z))$$

The gain of the new filter J at the frequency ω is then equal to that of H at a different frequency ϕ, chosen so that:

$$\cos(\phi) + i\sin(\phi) = R(\cos(\omega) + i\sin(\omega))$$

The function R moves points around on the unit circle; J at any point equals H on the point R moves it to.

For example, suppose we start with a one-zero, one-pole low-pass filter:

$$H(Z) = \frac{1 + Z^{-1}}{1 - gZ^{-1}}$$

and apply the function

$$R(Z) = -Z^2 = -\frac{1 \cdot Z^2 + 0 \cdot Z + 0}{0 \cdot Z^2 + 0 \cdot Z + 1}$$

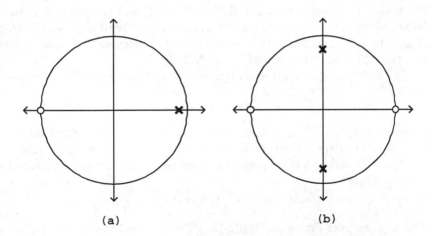

(a)   (b)

Figure 8.19: One-pole, one-zero low-pass filter: (a) pole-zero plot; (b) plot for the resulting filter after the transformation $R(Z) = -Z^2$. The result is a band-pass filter with center frequency $\pi/2$.

Geometrically, this choice of R stretches the unit circle uniformly to twice its circumference and wraps it around itself twice. The points 1 and -1 are both sent to the point -1, and the points i and $-i$ are sent to the point 1. The resulting transfer function is

$$J(Z) = \frac{1 - Z^{-2}}{1 + gZ^{-2}} = \frac{(1 - Z^{-1})(1 + Z^{-1})}{(1 - i\sqrt{g}Z^{-1})(1 + i\sqrt{g}Z^{-1})}$$

The pole-zero plots of H and J are shown in Figure 8.19. From a low-pass filter we ended up with a band-pass filter. The points i and $-i$ which R sends to 1 (where the original filter's gain is highest) become points of highest gain for the new filter.

8.3.8   Butterworth band-pass filter

We can apply the transformation $R(Z) = -Z^2$ to convert the Butterworth filter into a high-quality band-pass filter with center frequency $\pi/2$. A further transformation can then be applied to shift the center frequency to any desired value ω between 0 and π. The transformation will be of the form,

$$S(Z) = \frac{aZ + b}{bZ + a}$$

where a and b are real numbers and not both are zero. This is a particular case of the general form given above for unit-circle-preserving rational functions. We have $S(1) = 1$ and $S(-1) = -1$, and the top and bottom halves of the unit circle are transformed symmetrically (if Z goes to W then $\overline{Z}$ goes to $\overline{W}$). The qualitative effect of the transformation S is to squash points of the unit circle toward 1 or -1.

In particular, given a desired center frequency ω, we wish to choose S so that:

$$S(\cos(\omega) + i\sin(\omega)) = i$$

If we leave $R = -Z^2$ as before, and let H be the transfer function for a low-pass Butterworth filter, then the combined filter with transfer function $H(R(S(Z)))$ will be a band-pass filter with center frequency ω. Solving for a and b gives:

$$a = \cos(\frac{\pi}{4} - \frac{\omega}{2}), \ b = \sin(\frac{\pi}{4} - \frac{\omega}{2})$$

The new transfer function, $H(R(S(Z)))$, will have $2n$ poles and $2n$ zeros (if n is the degree of the Butterworth filter H).

Knowing the transfer function is good, but even better is knowing the locations of all the poles and zeros of the new filter, which we need to be able to compute it using elementary filters. If Z is a pole of the transfer

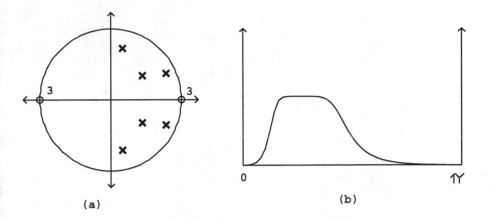

Figure 8.20: Butterworth band-pass filter: (a) pole-zero diagram; (b) frequency response. The center frequency is $\pi/4$. The bandwidth depends both on center frequency and on the bandwidth of the original Butterworth low-pass filter used.

function $J(Z) = H(R(S(Z)))$, that is, if $J(Z) = \infty$, then $R(S(Z))$ must be a pole of H. The same goes for zeros. To find a pole or zero of J we set $R(S(Z)) = W$, where W is a pole or zero of H, and solve for Z. This gives:

$$-\left[\frac{aZ+b}{bZ+a}\right]^2 = W$$

$$\frac{aZ+b}{bZ+a} = \pm\sqrt{-W}$$

$$Z = \frac{\pm a\sqrt{-W} - b}{\mp b\sqrt{-W} + a}$$

(Here a and b are as given above and we have used the fact that $a^2 + b^2 = 1$). A sample pole-zero plot and frequency response of J are shown in Figure 8.20.

8.3.9 Time-varying coefficients

In some recursive filter designs, changing the coefficients of the filter can inject energy into the system. A physical analogue is a child on a swing set. The child oscillates back and forth at the resonant frequency of the system, and pushing or pulling the child injects or extracts energy smoothly. However, if you decide to shorten the chain or move the swing set itself,

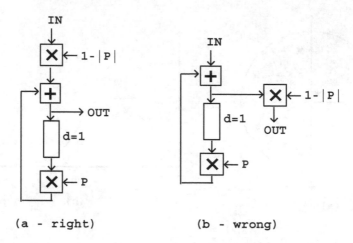

Figure 8.21: Normalizing a recirculating elementary filter: (a) correctly, by multiplying in the normalization factor at the input; (b) incorrectly, multiplying at the output.

you may inject an unpredictable amount of energy into the system. The same thing can happen when you change the coefficients in a resonant recirculating filter.

The simple one-zero and one-pole filters used here don't have this difficulty; if the feedback or feed-forward gain is changed smoothly (in the sense of an amplitude envelope) the output will behave smoothly as well. But one subtlety arises when trying to normalize a recursive filter's output when the feedback gain is close to one. For example, suppose we have a one-pole low-pass filter with gain 0.99 (for a cutoff frequency of 0.01 radians, or 70 Hertz at the usual sample rate). To normalize this for unit DC gain we multiply by 0.01. Suppose now we wish to double the cutoff frequency by changing the gain slightly to 0.98. This is fine except that the normalizing factor suddenly doubles. If we multiply the filter's output by the normalizing factor, the output will suddenly, although perhaps only momentarily, jump by a factor of two.

The trick is to normalize at the *input* of the filter, not the output. Figure 8.21 (part a) shows a complex recirculating filter, with feedback gain P, normalized at the input by $1 - |P|$ so that the peak gain is one. Part (b) shows the wrong way to do it, multiplying at the output.

Things get more complicated when several elementary recirculating filters are put in series, since the correct normalizing factor is in general a function of all the coefficients. One possible approach, if such a filter is required to change rapidly, is to normalize each input separately as if it were acting alone, then multiplying the output, finally, by whatever further correction is needed.

8.3.10   Impulse responses of recirculating filters

In Section 7.4 we analyzed the impulse response of a recirculating comb filter, of which the one-pole low-pass filter is a special case. Figure 8.22 shows the result for two low-pass filters and one complex one-pole resonant filter. All are elementary recirculating filters as introduced in Section 8.2.3. Each is normalized to have unit maximum gain.

In the case of a low-pass filter, the impulse response gets longer (and lower) as the pole gets closer to one. Suppose the pole is at a point $1 - 1/n$ (so that the cutoff frequency is $1/n$ radians). The normalizing factor is also $1/n$. After n points, the output diminishes by a factor of

$$\left(1 - \frac{1}{n}\right)^n \approx \frac{1}{e}$$

where e is Euler's constant, about 2.718. The filter can be said to have a *settling time* of n samples. In the figure, $n = 5$ for part (a) and $n = 10$ for part (b). In general, the settling time (in samples) is approximately one over the cutoff frequency (in angular units).

The situation gets more interesting when we look at a resonant one-pole filter, that is, one whose pole lies off the real axis. In part (c) of the figure, the pole P has absolute value 0.9 (as in part b), but its argument is set to $2\pi/10$ radians. We get the same settling time as in part (b), but the output rings at the resonant frequency (and so at a period of 10 samples in this example).

A natural question to ask is, how many periods of ringing do we get before the filter decays to strength $1/e$? If the pole of a resonant filter has magnitude $1 - 1/n$ as above, we have seen in Section 8.2.3 that the bandwidth (call it b) is about $1/n$, and we see here that the settling time is about n. The resonant frequency (call it ω) is the argument of the pole, and the period in samples of the ringing is $2\pi/\omega$. The number of periods that make up the settling time is thus:

$$\frac{n}{2\pi/\omega} = \frac{1}{2\pi}\frac{\omega}{b} = \frac{q}{2\pi}$$

where q is the *quality* of the filter, defined as the center frequency divided by bandwidth. Resonant filters are often specified in terms of the center frequency and "q" in place of bandwidth.

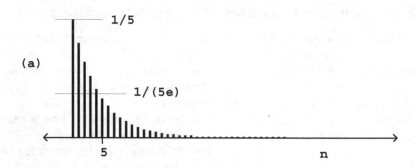

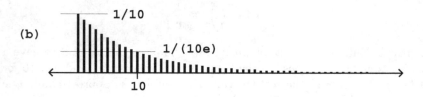

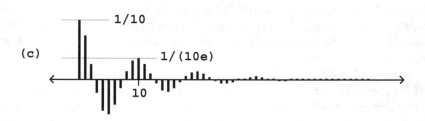

Figure 8.22: The impulse response of three elementary recirculating (one-pole) filters, normalized for peak gain 1: (a) low-pass with $P = 0.8$; (b) low-pass with $P = 0.9$; (c) band-pass (only the real part shown), with $|P| = 0.9$ and a center frequency of $2\pi/10$.

8.3.11 All-pass filters

Sometimes a filter is applied to get a desired phase change, rather than to alter the amplitudes of the frequency components of a sound. To do this we would need a way to design a filter with a constant, unit frequency response but which changes the phase of an incoming sinusoid in a way that depends on its frequency. We have already seen in Chapter 7 that a delay of length d introduces a phase change of $-d\omega$, at the angular frequency ω. Another class of filters, called *all-pass filters*, can make phase changes which are more interesting functions of ω.

To design an all-pass filter, we start with two facts: first, an elementary recirculating filter and an elementary non-recirculating one cancel each other out perfectly if they have the same gain coefficient. In other words, if a signal has been put through a one-zero filter, either real or complex, the effect can be reversed by sequentially applying a one-pole filter, and vice versa.

The second fact is that the elementary non-recirculating filter of the second form has the same frequency response as that of the first form; they differ only in phase response. So if we combine an elementary recirculating filter with an elementary non-recirculating one of the second form, the frequency responses cancel out (to a flat gain independent of frequency) but the phase response is not constant.

To find the transfer function, we choose the same complex number $P < 1$ as coefficient for both elementary filters and multiply their transfer

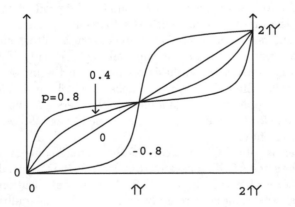

Figure 8.23: Phase response of all-pass filters with different pole locations p. When the pole is located at zero, the filter reduces to a one-sample delay.

functions:

$$H(Z) = \frac{\overline{P} - Z^{-1}}{1 - PZ^{-1}}$$

The coefficient P controls both the location of the one pole (at P itself) and the zero (at $1/\overline{P}$). Figure 8.23 shows the phase response of the all-pass filter for four real-valued choices p of the coefficient. At frequencies of 0, π, and 2π, the phase response is just that of a one-sample delay; but for frequencies in between, the phase response is bent upward or downward depending on the coefficient.

Complex coefficients give similar phase response curves, but the frequencies at which they cross the diagonal line in the figure are shifted according to the argument of the coefficient P.

8.4   Applications

Filters are used in a broad range of applications both in audio engineering and in electronic music. The former include, for instance, equalizers, speaker crossovers, sample rate converters, and DC removal (which we have already used in earlier chapters). Here, though, we'll be concerned with the specifically musical applications.

8.4.1   Subtractive synthesis

Subtractive synthesis is the technique of using filters to shape the spectral envelope of a sound, forming another sound, usually preserving qualities of the original sound such as pitch, roughness, noisiness, or graniness. The spectral envelope of the resulting sound is the product of the spectral envelope of the original sound with the frequency response of the filter. Figure 8.24 shows a possible configuration of source, filter, and result.

The filter may be constant or time-varying. Already in wide use by the mid 1950s, subtractive synthesis boomed with the introduction of the voltage-controlled filter (VCF), which became widely available in the mid 1960s with the appearance of modular synthesizers. A typical VCF has two inputs: one for the sound to filter, and one to vary the center or cutoff frequency of the filter.

A popular use of a VCF is to control the center frequency of a resonant filter from the same ADSR generator that controls the amplitude; a possible block diagram is shown in Figure 8.25. In this configuration, the louder portion of a note (loudness roughly controlled by the multiplier at the bottom) may also be made to sound brighter, using the filter, than the quieter parts; this can mimic the spectral evolution of strings or brass instruments over the life of a note.

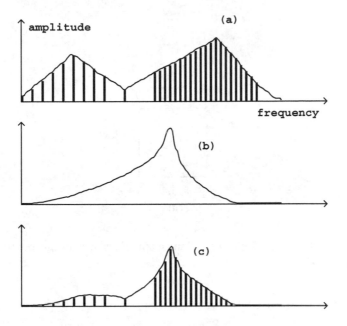

Figure 8.24: Subtractive synthesis: (a) spectrum of input sound; (b) filter frequency response; (c) spectrum of output sound.

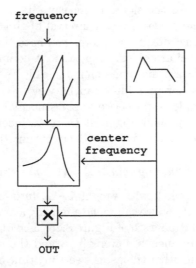

Figure 8.25: ADSR-controlled subtractive synthesis.

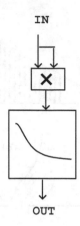

IN

OUT

Figure 8.26: Envelope follower. The output is the average power of the input signal.

8.4.2 Envelope following

It is frequently desirable to use the time-varying power of an incoming signal to trigger or control a musical process. To do this, we will need a procedure for measuring the power of an audio signal. Since most audio signals pass through zero many times per second, it won't suffice to take instantaneous values of the signal to measure its power; instead, we must calculate the average power over an interval of time long enough that its variations won't show up in the power estimate, but short enough that changes in signal level are quickly reported. A computation that provides a time-varying power estimate of a signal is called an *envelope follower*.

The output of a low-pass filter can be viewed as a moving average of its input. For example, suppose we apply a normalized one-pole low-pass filter with coefficient p, as in Figure 8.21, to an incoming signal $x[n]$. The output (call it y[n]) is the sum of the delay output times p, with the input times $1 - p$:

$$y[n] = p \cdot y[n-1] + (1-p) \cdot x[n]$$

so each input is averaged, with weight $1 - p$, into the previous output to produce a new output. So we can make a moving average of the square of an audio signal using the diagram of Figure 8.26. The output is a time-varying average of the instantaneous power $x[n]^2$, and the design of the low-pass filter controls, among other things, the settling time of the moving average.

For more insight into the design of a suitable low-pass filter for an envelope follower, we analyze it from the point of view of signal spectra. If,

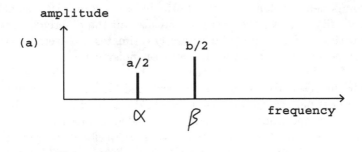

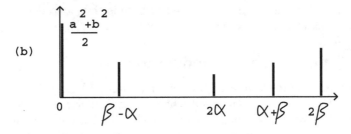

Figure 8.27: Envelope following from the spectral point of view: (a) an incoming signal with two components; (b) the result of squaring it.

for instance, we put in a real-valued sinusoid:

$$x[n] = a \cdot \cos(\alpha n)$$

the result of squaring is:

$$x[n]^2 = \frac{a^2}{2} (\cos(2\alpha n) + 1)$$

and so if the low-pass filter effectively stops the component of frequency 2α we will get out approximately the constant $a^2/2$, which is indeed the average power.

The situation for a signal with several components is similar. Suppose the input signal is now,

$$x[n] = a \cdot \cos(\alpha n) + b \cdot \cos(\beta n)$$

whose spectrum is plotted in Figure 8.27 (part a). (We have omitted the two phase terms but they will have no effect on the outcome.) Squaring the signal produces the spectrum shown in part (b) (see Section 5.2).) We can get the desired fixed value of $(a^2 + b^2)/2$ simply by filtering out all the other components; ideally the result will be a constant (DC) signal. As long as we filter out all the partials, and also all the difference tones, we end up with a stable output that correctly estimates the average power.

Envelope followers may also be used on noisy signals, which may be thought of as signals with dense spectra. In this situation there will be difference frequencies arbitrarily close to zero, and filtering them out entirely will be impossible; we will always get fluctuations in the output, but they will decrease proportionally as the filter's passband is narrowed.

Although a narrower passband will always give a cleaner output, whether for discrete or continuous spectra, the filter's settling time will lengthen proportionally as the passband is narrowed. There is thus a tradeoff between getting a quick response and a smooth result.

8.4.3　Single sideband modulation

As we saw in Chapter 5, multiplying two real sinusoids together results in a signal with two new components at the sum and difference of the original frequencies. If we carry out the same operation with complex sinusoids, we get only one new resultant frequency; this is one result of the greater mathematical simplicity of complex sinusoids as compared to real ones. If we multiply a complex sinusoid $1, Z, Z^2, \ldots$ with another one, $1, W, W^2, \ldots$ the result is $1, WZ, (WZ)^2, \ldots$, which is another complex sinusoid whose frequency, $\angle(ZW)$, is the sum of the two original frequencies.

In general, since complex sinusoids have simpler properties than real ones, it is often useful to be able to convert from real sinusoids to complex ones. In other words, from the real sinusoid:

$$x[n] = a \cdot \cos(\omega n)$$

(with a spectral peak of amplitude $a/2$ and frequency ω) we would like a way of computing the complex sinusoid:

$$X[n] = a\left(\cos(\omega n) + i\sin(\omega n)\right)$$

so that

$$x[n] = \mathrm{re}(X[n])$$

We would like a linear process for doing this, so that superpositions of sinusoids get treated as if their components were dealt with separately.

Of course we could equally well have chosen the complex sinusoid with frequency $-\omega$:

$$X'[n] = a\left(\cos(\omega n) - i\sin(\omega n)\right)$$

and in fact $x[n]$ is just half the sum of the two. In essence we need a filter that will pass through positive frequencies (actually frequencies between 0 and π, corresponding to values of Z on the top half of the complex unit circle) from negative values (from $-\pi$ to 0, or equivalently, from π to 2π— the bottom half of the unit circle).

One can design such a filter by designing a low-pass filter with cutoff frequency $\pi/2$, and then performing a rotation by $\pi/2$ radians using the technique of Section 8.3.4. However, it turns out to be easier to do it using two specially designed networks of all-pass filters with real coefficients.

Calling the transfer functions of the two filters H_1 and H_2, we design the filters so that

$$\angle(H_1(Z)) - \angle(H_2(Z)) \approx \begin{cases} \pi/2 & 0 < \angle(Z) < \pi \\ -\pi/2 & -\pi < \angle(Z) < 0 \end{cases}$$

or in other words,

$$H_1(Z) \approx iH_2(Z), \; 0 < \angle(Z) < \pi$$

$$H_1(Z) \approx -iH_2(Z), \; -\pi < \angle(Z) < 0$$

Then for any incoming real-valued signal $x[n]$ we simply form a complex number $a[n] + ib[n]$ where $a[n]$ is the output of the first filter and $b[n]$ is the output of the second. Any complex sinusoidal component of $x[n]$ (call it Z^n) will be transformed to

$$H_1(Z) + iH_2(Z) \approx \begin{cases} 2H_1(Z) & 0 < \angle(Z) < \pi \\ 0 & \text{otherwise} \end{cases}$$

Having started with a real-valued signal, whose energy is split equally into positive and negative frequencies, we end up with a complex-valued one with only positive frequencies.

8.5 Examples

In this section we will first introduce some easy-to-use prefabricated filters available in Pd to develop examples showing applications from the previous section. Then we will show some more sophisticated applications that require specially designed filters.

Prefabricated low-, high-, and band-pass filters

Patches H01.low-pass.pd, H02.high-pass.pd, and H03.band-pass.pd (Figure 8.28) show Pd's built-in filters, which implement filter designs described in Sections 8.3.1, 8.3.2 and 8.3.4. Two of the patches also use a noise generator we have not introduced before. We will need four new Pd objects:

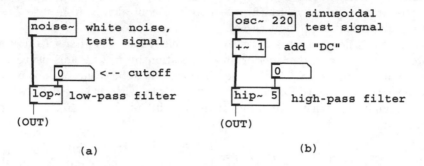

Figure 8.28: Using prefabricated filters in Pd: (a) a low-pass filter, with white noise as a test input; (b) using a high-pass filter to remove a signal component of frequency 0.

lop˜ : one-pole low-pass filter. The left inlet takes a signal to be filtered, and the right inlet takes control messages to set the cutoff frequency of the filter. The filter is normalized so that the gain is one at frequency 0.

hip˜ : one-pole, one-zero high-pass filter, with the same inputs and outputs as lop˜, normalized to have a gain of one at the Nyquist frequency.

bp˜ : resonant filter. The middle inlet takes control messages to set the center frequency, and the right inlet to set "q".

noise˜ : white noise generator. Each sample is an independent pseudo-random number, uniformly distributed from -1 to 1.

The first three example patches demonstrate these three filters (see Figure 8.28). The lop˜ and bp˜ objects are demonstrated with noise as input; hip˜ as shown is used to remove the DC (zero frequency) component of a signal.

Prefabricated time-varying band-pass filter

Time-varying band-pass filtering, as often used in classical subtractive synthesis (Section 8.4.1), can be done using the vcf˜ object, introduced here:

vcf˜ : a "voltage controlled" band-pass filter, similar to bp˜, but with a signal inlet to control center frequency. Both bp˜ and vcf˜ are one-pole resonant filters as developed in Section 8.3.4; bp˜ outputs only the real part of the resulting signal, while vcf˜ outputs the real and imaginary parts separately.

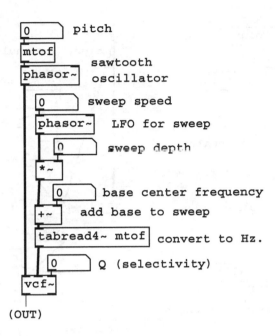

Figure 8.29: The vcf~ band-pass filter, with its center frequency controlled by an audio signal (as compared to bp~ which takes only control messages to set its center frequency.

Example H04.filter.sweep.pd (Figure 8.29) demonstrates using the vcf~ object for a simple and characteristic subtractive synthesis task. A phasor~ object (at top) creates a sawtooth wave to filter. (This is not especially good practice as we are not controlling the possibility of foldover; a better sawtooth generator for this purpose will be developed in Chapter 10.) The second phasor~ object (labeled "LFO for sweep") controls the time-varying center frequency. After adjusting to set the depth and a base center frequency (given in MIDI units), the result is converted into Hertz (using the tabread4~ object) and passed to vcf~ to set its center frequency. Another example of using a vcf~ object for subtractive synthesis is demonstrated in example H05.filter.floyd.pd.

Envelope followers

Example H06.envelope.follower.pd shows a simple and self-explanatory realization of the envelope follower described in Section 8.4.2. An interesting application of envelope following is shown in Example

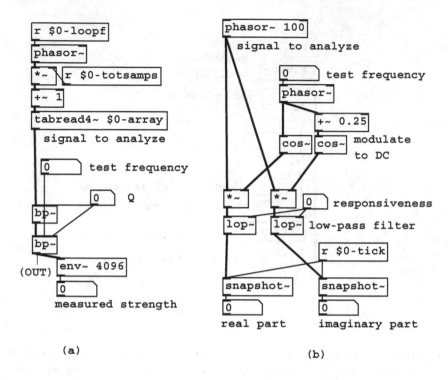

(a)   (b)

Figure 8.30: Analyzing the spectrum of a sound: (a) band-pass filtering a sampled bell sound and envelope-following the result; (b) frequency-shifting a partial to DC and reading off its real and imaginary part.

H07.measure.spectrum.pd (Figure 8.30, part a). A famous bell sample is looped as a test sound. Rather than get the overall mean square power of the bell, we would like to estimate the frequency and power of each of its partials. To do this we sweep a band-pass filter up and down in frequency, listening to the result and/or watching the filter's output power using an envelope follower. (We use two band-pass filters in series for better isolation of the partials; this is not especially good filter design practice but it will do in this context.) When the filter is tuned to a partial the envelope follower reports its strength.

Example H08.heterodyning.pd (part (b) of the figure) shows an alternative way of finding partial strengths of an incoming sound; it has the advantage of reporting the phase as well as the strength. First we modulate the desired partial down to zero frequency. We use a complex-valued sinusoid as a modulator so that we get only one sideband for each component of the

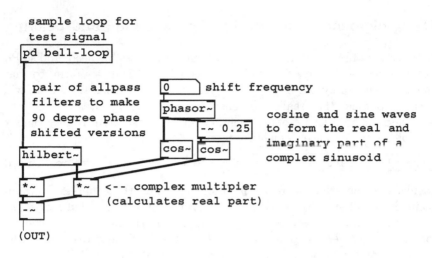

Figure 8.31: Using an all-pass filter network to make a frequency shifter.

input. The test frequency is the only frequency that is modulated to DC; others go elsewhere. We then low-pass the resulting complex signal. (We can use a real-valued low-pass filter separately on the real and imaginary parts.) This essentially removes all the partials except for the DC one, which we then harvest. This technique is the basis of Fourier analysis, the subject of Chapter 9.

Single sideband modulation

As described in Section 8.4.3, a pair of all-pass filters can be constructed to give roughly $\pi/2$ phase difference for positive frequencies and $-\pi/2$ for negative ones. The design of these pairs is beyond the scope of this discussion (see, for instance, [Reg93]) but Pd does provide an abstraction, hilbert~, to do this. Example H09.ssb.modulation.pd, shown in Figure 8.31, demonstrates how to use the hilbert~ abstraction to do signal sideband modulation. The Hilbert transform dates to the analog era [Str95, pp.129-132].

The two outputs of hilbert~, considered as the real and imaginary parts of a complex-valued signal, are multiplied by a complex sinusoid (at right in the figure), and the real part is output. The components of the resulting signal are those of the input shifted by a (positive or negative) frequency specified in the number box.

Using elementary filters directly: shelving and peaking

No finite set of prefabricated filters could fill every possible need, and so Pd provides the elementary filters of Sections 8.2.1-8.2.3 in raw form, so that the user can supply the filter coefficients explicitly. In this section we will describe patches that realize the shelving and peaking filters of Sections 8.3.3 and 8.3.5 directly from elementary filters. First we introduce the six Pd objects that realize elementary filters:

| rzero~ |, | rzero_rev~ |, | rpole~ |: elementary filters with real-valued

coefficients operating on real-valued signals. The three implement non-recirculating filters of the first and second types, and the recirculating filter. They all have one inlet, at right, to supply the coefficient that sets the location of the zero or pole. The inlet for the coefficient (as well as the left inlet for the signal to filter) take audio signals. No stability check is performed.

| czero~ |, | czero_rev~ |, | cpole~ |: elementary filters with complex-valued

coefficients, operating on complex-valued signals, corresponding to the real-valued ones above. Instead of two inlets and one outlet, each of these filters has four inlets (real and imaginary part of the signal to filter, and real and imaginary part of the coefficient) and two outlets for the complex-valued output.

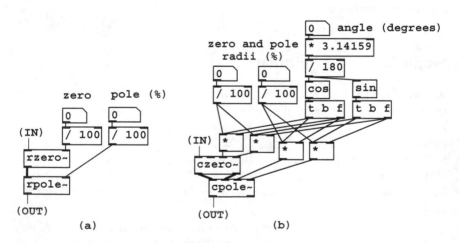

Figure 8.32: Building filters from elementary, raw ones: (a) shelving; (b) peaking.

The example patches use a pair of abstractions to graph the frequency and phase responses of filters as explained in Example H10.measurement.pd. Example H11.shelving.pd (Figure 8.32, part a) shows how to make a shelving filter. One elementary non-recirculating filter (**rzero~**) and one elementary recirculating one (**rpole~**) are put in series. As the analysis of Section 8.3.9 might suggest, the **rzero~** object is placed first.

Example H12.peaking.pd (part (b) of the figure) implements a peaking filter. Here the pole and the zero are rotated by an angle ω to control the center frequency of the filter. The bandwidth and center frequency gain are equal to the shelf frequency and the DC gain of the corresponding shelving filter.

Example H13.butterworth.pd demonstrates a three-pole, three-zero Butterworth shelving filter. The filter itself is an abstraction, **butterworth3~**, for easy reuse.

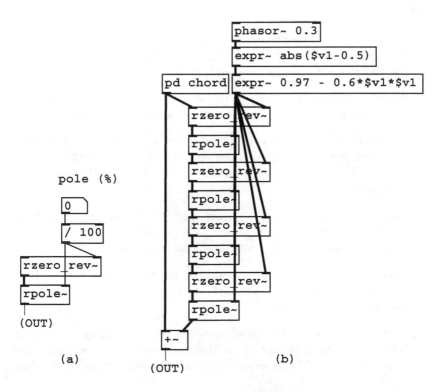

Figure 8.33: All-pass filters: (a) making an all-pass filter from elementary filters; (b) using four all-pass filters to build a phaser.

Making and using all-pass filters

Example H14.all.pass.pd (Figure 8.33, part a) shows how to make an all-pass filter out of a non-recirculating filter, second form (`rzero_rev~`) and a recirculating filter (`rpole~`). The coefficient, ranging from -1 to 1, is controlled in hundredths.

Example H15.phaser.pd (part b of the figure) shows how to use four all-pass filters to make a classic phaser. The phaser works by summing the input signal with a phase-altered version of it, making interference effects. The amount of phase change is varied in time by varying the (shared) coefficient of the all-pass filters. The overall effect is somewhat similar to a flanger (time-varying comb filter) but the phaser does not impose a pitch as the comb filter does.

Exercises

1. A recirculating elementary filter has a pole at $i/2$. At what angular frequency is its gain greatest, and what is the gain there? At what angular frequency is the gain least, and what is the gain there?

2. A shelving filter has a pole at 0.9 and a zero at 0.8. What are: the DC gain; the gain at Nyquist; the approximate transition frequency?

3. Suppose a complex recirculating filter has a pole at P. Suppose further that you want to combine its real and imaginary output to make a single, real-valued signal equivalent to a two-pole filter with poles at P and $\overline{P}$. How would you weight the two outputs?

4. Suppose you wish to design a peaking filter with gain 2 at 1000 Hertz and bandwidth 200 Hertz (at a sample rate of 44100 Hertz). Where, approximately, would you put the upper pole and zero?

5. In the same situation, where would you put the (upper) pole and zero to remove a sinusoid at 1000 Hertz entirely, while attenuating only 3 decibels at 1001 Hertz?

6. A one-pole complex filter is excited by an impulse to make a tone at 1000 Hertz, which decays 10 decibels in one second (at a sample rate of 44100 Hertz). Where would you place the pole? What is the value of "q"?

Chapter 9

Fourier Analysis and Resynthesis

Among the applications of filters discussed in Chapter 8, we saw how to use heterodyning, combined with a low-pass filter, to find the amplitude and phase of a sinusoidal component of a signal (Page 261). In this chapter we will refine this technique into what is called *Fourier analysis*. In its simplest form, Fourier analysis takes as input any periodic signal (of period N) and outputs the complex-valued amplitudes of its N possible sinusoidal components. These N complex amplitudes can theoretically be used to reconstruct the original signal exactly. This reconstruction is called *Fourier resynthesis*.

In this chapter we will start by developing the theory of Fourier analysis and resynthesis of periodic sampled signals. Then we will go on to show how to apply the same techniques to arbitrary signals, whether periodic or not. Finally, we will develop some standard applications such as the phase vocoder.

9.1 Fourier Analysis of Periodic Signals

Suppose $X[n]$ is a complex-valued signal that repeats every N samples. (We are continuing to use complex-valued signals rather than real-valued ones to simplify the mathematics.) Because of the period N, the values of $X[n]$ for $n = 0, \ldots, N-1$ determine $X[n]$ for all integer values of n.

Suppose further that $X[n]$ can be written as a sum of complex sinusoids of frequency 0, $2\pi/N$, $4\pi/N$, ..., $2(N-1)\pi/N$. These are the partials, starting with the zeroth, for a signal of period N. We stop at the Nth term

267

because the next one would have frequency 2π, equivalent to frequency 0, which is already on the list.

Given the values of X, we wish to find the complex amplitudes of the partials. Suppose we want the kth partial, where $0 \le k < N$. The frequency of this partial is $2\pi k/N$. We can find its complex amplitude by modulating X downward $2\pi k/N$ radians per sample in frequency, so that the kth partial is modulated to frequency zero. Then we pass the signal through a low-pass filter with such a low cutoff frequency that nothing but the zero-frequency partial remains. We can do this in effect by averaging over a huge number of samples; but since the signal repeats every N samples, this huge average is the same as the average of the first N samples. In short, to measure a sinusoidal component of a periodic signal, modulate it down to DC and then average over one period.

Let $\omega = 2\pi/N$ be the fundamental frequency for the period N, and let U be the unit-magnitude complex number with argument ω:

$$U = \cos(\omega) + i \sin(\omega)$$

The kth partial of the signal $X[n]$ is of the form:

$$P_k[n] = A_k \left[U^k\right]^n$$

where A_k is the complex amplitude of the partial, and the frequency of the partial is:

$$\angle(U^k) = k\angle(U) = k\omega$$

We're assuming for the moment that the signal $X[n]$ can actually be written as a sum of the n partials, or in other words:

$$X[n] = A_0 \left[U^0\right]^n + A_1 \left[U^1\right]^n + \cdots + A_{N-1}\left[U^{N-1}\right]^n$$

By the heterodyne-filtering argument above, we expect to be able to measure each A_k by multiplying by the sinusoid of frequency $-k\omega$ and averaging over a period:

$$A_k = \frac{1}{N}\left(\left[U^{-k}\right]^0 X[0] + \left[U^{-k}\right]^1 X[1] + \cdots + \left[U^{-k}\right]^{N-1} X[N-1]\right)$$

This is such a useful formula that it gets its own notation. The *Fourier transform* of a signal $X[n]$, over N samples, is defined as:

$$\mathcal{FT}\left\{X[n]\right\}(k) = V^0 X[0] + V^1 X[1] + \cdots + V^{N-1} X[N-1]$$

where $V = U^{-k}$. The Fourier transform is a function of the variable k, equal to N times the amplitude of the input's kth partial. So far k has taken integer values but the formula makes sense for any value of k if we define V more generally as:

$$V = \cos(-k\omega) + i\sin(-k\omega)$$

where, as before, $\omega = 2\pi/N$ is the (angular) fundamental frequency associated with the period N.

9.1.1 Periodicity of the Fourier transform

If X[n] is, as above, a signal that repeats every N samples, the Fourier transform of X[n] also repeats itself every N units of frequency, that is,

$$\mathcal{FT}\{X[n]\}(k+N) = \mathcal{FT}\{X[n]\}(k)$$

for all real values of k. This follows immediately from the definition of the Fourier transform, since the factor

$$V = \cos(-k\omega) + i\sin(-k\omega)$$

is unchanged when we add N (or any multiple of N) to k.

9.1.2 Fourier transform as additive synthesis

Now consider an arbitrary signal $X[n]$ that repeats every N samples. (Previously we had assumed that $X[n]$ could be obtained as a sum of sinusoids, and we haven't yet found out whether every periodic $X[n]$ can be obtained that way.) Let $Y[k]$ denote the Fourier transform of X for $k = 0, ..., N-1$:

$$Y[k] = \mathcal{FT}\{X[n]\}(k)$$

$$= \left[U^{-k}\right]^0 X[0] + \left[U^{-k}\right]^1 X[1] + \cdots + \left[U^{-k}\right]^{N-1} X[N-1]$$

$$= \left[U^0\right]^k X[0] + \left[U^{-1}\right]^k X[1] + \cdots + \left[U^{-(N-1)}\right]^k X[N-1]$$

In the second version we rearranged the exponents to show that $Y[k]$ is a sum of complex sinusoids, with complex amplitudes $X[m]$ and frequencies $-m\omega$ for $m = 0, \ldots, N-1$. In other words, $Y[k]$ can be considered as a Fourier series in its own right, whose mth component has strength $X[-m]$. (The expression $X[-m]$ makes sense because X is a periodic signal). We can also express the amplitude of the partials of $Y[k]$ in terms of its own Fourier transform. Equating the two gives:

$$\frac{1}{N}\mathcal{FT}\{Y[k]\}(m) = X[-m]$$

This means in turn that $X[-m]$ can be obtained by summing sinusoids with amplitudes $Y[k]/N$. Setting $n = -m$ gives:

$$X[n] = \frac{1}{N} \mathcal{FT}\left\{Y[k]\right\}(-n)$$

$$= \left[U^0\right]^n Y[0] + \left[U^1\right]^n Y[1] + \cdots + \left[U^{N-1}\right]^n Y[N-1]$$

This shows that any periodic $X[n]$ can indeed be obtained as a sum of sinusoids. Further, the formula explicitly shows how to reconstruct $X[n]$ from its Fourier transform $Y[k]$, if we know its value for the integers $k = 0, \ldots, N-1$.

9.2   Properties of Fourier Transforms

In this section we will investigate what happens when we take the Fourier transform of a (complex) sinusoid. The simplest one is "DC", the special sinusoid of frequency zero. After we derive the Fourier transform of that, we will develop some properties of Fourier transforms that allow us to apply the result to any other sinusoid.

9.2.1   Fourier transform of DC

Let $X[n] = 1$ for all n (this repeats with any desired integer period $N > 1$). From the preceding discussion, we expect to find that

$$\mathcal{FT}\left\{X[n]\right\}(k) = \begin{cases} N & k = 0 \\ 0 & k = 1, \ldots, N-1 \end{cases}$$

We will often need to know the answer for non-integer values of k however, and for this there is nothing better to do than to calculate the value directly:

$$\mathcal{FT}\left\{X[n]\right\}(k) = V^0 X[0] + V^1 X[1] + \cdots + V^{N-1} X[N-1]$$

where V is, as before, the unit magnitude complex number with argument $-k\omega$. This is a geometric series; as long as $V \neq 1$ we get:

$$\mathcal{FT}\left\{X[n]\right\}(k) = \frac{V^N - 1}{V - 1}$$

We now symmetrize the top and bottom in the same way as we earlier did in Section 7.3. To do this let:

$$\xi = \cos(\pi k/N) - i\sin(\pi k/N)$$

so that $\xi^2 = V$. Then factoring appropriate powers of ξ out of the numerator and denominator gives:

$$\mathcal{FT}\left\{X[n]\right\}(k) = \xi^{N-1}\frac{\xi^N - \xi^{-N}}{\xi - \xi^{-1}}$$

It's easy now to simplify the numerator:

$$\xi^N - \xi^{-N} = (\cos(\pi k) - i\sin(\pi k)) - (\cos(\pi k) + i\sin(\pi k)) = -2i\sin(\pi k)$$

and similarly for the denominator, giving:

$$\mathcal{FT}\left\{X[n]\right\}(k) = \Big(\cos(\pi k(N-1)/N) - i\sin(\pi k(N-1)/N)\Big)\frac{\sin(\pi k)}{\sin(\pi k/N)}$$

Whether $V = 1$ or not, we have

$$\mathcal{FT}\left\{X[n]\right\}(k) = \Big(\cos(\pi k(N-1)/N) - i\sin(\pi k(N-1)/N)\Big)D_N(k)$$

where $D_N(k)$, known as the *Dirichlet kernel*, is defined as

$$D_N(k) = \begin{cases} N & k = 0 \\ \frac{\sin(\pi k)}{\sin(\pi k/N)} & k \neq 0,\ -N < k < N \end{cases}$$

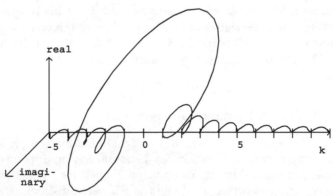

Figure 9.1: The Fourier transform of a signal consisting of all ones. Here N=100, and values are shown for k ranging from -5 to 10. The result is complex-valued and shown as a projection, with the real axis pointing up the page and the imaginary axis pointing away from it.

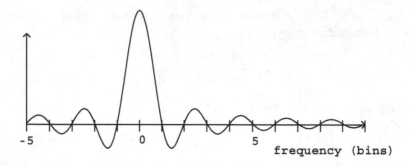

Figure 9.2: The Dirichlet kernel, for $N = 100$.

Figure 9.1 shows the Fourier transform of $X[n] = 1$, with $N = 100$. The transform repeats every 100 samples, with a peak at $k = 0$, another at $k = 100$, and so on. The figure endeavors to show both the magnitude and phase behavior using a 3-dimensional graph projected onto the page. The phase term

$$\cos(\pi k(N-1)/N) - i\sin(\pi k(N-1)/N)$$

acts to twist the values of $\mathcal{FT}\{X[n]\}(k)$ around the k axis with a period of approximately two. The Dirichlet kernel $D_N(k)$, shown in Figure 9.2, controls the magnitude of $\mathcal{FT}\{X[n]\}(k)$. It has a peak, two units wide, around $k = 0$. This is surrounded by one-unit-wide *sidelobes*, alternating in sign and gradually decreasing in magnitude as k increases or decreases away from zero. The phase term rotates by almost π radians each time the Dirichlet kernel changes sign, so that the product of the two stays roughly in the same complex half-plane for $k > 1$ (and in the opposite half-plane for $k < -1$). The phase rotates by almost 2π radians over the peak from $k = -1$ to $k = 1$.

9.2.2   Shifts and phase changes

Section 7.2 showed how time-shifting a signal changes the phases of its sinusoidal components, and Section 8.4.3 showed how multiplying a signal by a complex sinusoid shifts its component frequencies. These two effects have corresponding identities involving the Fourier transform.

First we consider a time shift. If $X[n]$, as usual, is a complex-valued signal that repeats every N samples, let $Y[n]$ be $X[n]$ delayed d samples:

$$Y[n] = X[n - d]$$

which also repeats every N samples since X does. We can reduce the Fourier transform of $Y[n]$ this way:

$$\mathcal{FT}\{Y[n]\}(k) = V^0 Y[0] + V^1 Y[1] + \cdots + V^{N-1} Y[N-1]$$

$$= V^0 X[-d] + V^1 X[-d+1] + \cdots + V^{N-1} X[-d+N-1]$$

$$= V^d X[0] + V^{d+1} X[1] + \cdots + V^{d+N-1} X[N-1]$$

$$= V^d \left(V^0 X[0] + V^1 X[1] + \cdots + V^{N-1} X[N-1] \right)$$

$$= V^d \mathcal{FT}\{X[n]\}(k)$$

(The third line is just the second one with the terms summed in a different order). We therefore get the Time Shift Formula for Fourier Transforms:

$$\mathcal{FT}\{X[n-d]\}(k) = \Big(\cos(-dk\omega) + i \sin(-dk\omega) \Big) \mathcal{FT}\{X[n]\}(k)$$

The Fourier transform of $X[n-d]$ is a phase term times the Fourier transform of $X[n]$. The phase is changed by $-dk\omega$, a linear function of the frequency k.

Now suppose instead that we change our starting signal $X[n]$ by multiplying it by a complex exponential Z^n with angular frequency α:

$$Y[n] = Z^n X[n]$$

$$Z = \cos(\alpha) + i \sin(\alpha)$$

The Fourier transform is:

$$\mathcal{FT}\{Y[n]\}(k) = V^0 Y[0] + V^1 Y[1] + \cdots + V^{N-1} Y[N-1]$$

$$= V^0 X[0] + V^1 Z X[1] + \cdots + V^{N-1} Z^{N-1} X[N-1]$$

$$= (VZ)^0 X[0] + (VZ)^1 X[1] + \cdots + (VZ)^{N-1} X[N-1]$$

$$= \mathcal{FT}\{X[n]\}\left(k - \frac{\alpha}{\omega}\right)$$

We therefore get the Phase Shift Formula for Fourier Transforms:

$$\mathcal{FT}\{(\cos(\alpha) + i \sin(\alpha)) X[n]\}(k) = \mathcal{FT}\{X[n]\}\left(k - \frac{\alpha N}{2\pi}\right)$$

9.2.3   Fourier transform of a sinusoid

We can use the phase shift formula above to find the Fourier transform of any complex sinusoid Z^n with frequency α, simply by setting $X[n] = 1$ in the formula and using the Fourier transform for DC:

$$\mathcal{FT}\{Z^n\}(k) = \mathcal{FT}\{1\}(k - \frac{\alpha}{\omega})$$

$$= [\cos(\Phi(k)) + i\sin(\Phi(k))]\, D_N(k - \frac{\alpha}{\omega})$$

where D_N is the Dirichlet kernel and Φ is an ugly phase term:

$$\Phi(k) = -\pi \cdot (k - \frac{\alpha}{\omega}) \cdot (N - 1)/N$$

If the sinusoid's frequency α is an integer multiple of the fundamental frequency ω, the Dirichlet kernel is shifted to the left or right by an integer. In this case the zero crossings of the Dirichlet kernel line up with integer values of k, so that only one partial is nonzero. This is pictured in Figure 9.3 (part a).

Part (b) shows the result when the frequency α falls halfway between two integers. The partials have amplitudes falling off roughly as $1/k$ in both directions, measured from the actual frequency α. That the energy should be spread over many partials, when after all we started with a single sinusoid, might seem surprising at first. However, as shown in Figure 9.4, the signal repeats at a period N which disagrees with the frequency of the sinusoid. As a result there is a discontinuity at the beginning of each period, and energy is flung over a wide range of frequencies.

9.3   Fourier Analysis of Non-Periodic Signals

Most signals aren't periodic, and even a periodic one might have an unknown period. So we should be prepared to do Fourier analysis on signals without making the comforting assumption that the signal to analyze repeats at a fixed period N. Of course, we can simply take N samples of the signal and *make* it periodic; this is essentially what we did in the previous section, in which a pure sinusoid gave us the complicated Fourier transform of Figure 9.3 (part b).

However, it would be better to get a result in which the response to a pure sinusoid were better localized around the corresponding value of k. We can accomplish this using the enveloping technique first introduced in Figure 2.7 (Page 38). Applying this technique to Fourier analysis will not

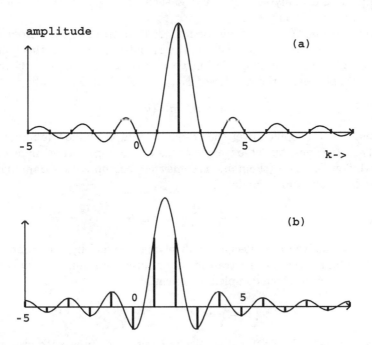

Figure 9.3: Fourier transforms of complex sinusoids, with $N = 100$: (a) with frequency 2ω ; (b) with frequency 1.5ω. (The effect of the phase winding term is not shown.)

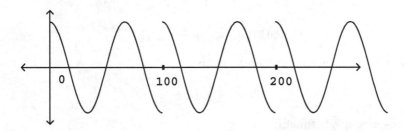

Figure 9.4: A complex sinusoid with frequency $\alpha = 1.5\omega = 3\pi/N$, forced to repeat every N samples. (N is arbitrarily set to 100; only the real part is shown.)

only improve our analyses, but will also shed new light on the enveloping looping sampler of Chapter 2.

Given a signal $X[n]$, periodic or not, defined on the points from 0 to $N - 1$, the technique is to envelope the signal before doing the Fourier analysis. The envelope shape is known as a *window function*. Given a window function $w[n]$, the *windowed Fourier transform* is:

$$\mathcal{FT}\left\{w[n]X[n]\right\}(k)$$

Much ink has been spilled over the design of suitable window functions for particular situations, but here we will consider the simplest one, named the *Hann* window function (the name is sometimes corrupted to "Hanning" in DSP circles). The Hann window is:

$$w[n] = \frac{1}{2} - \frac{1}{2}\cos(2\pi n/N)$$

It is easy to analyze the effect of multiplying a signal by the Hann window before taking the Fourier transform, because the Hann window can be written as a sum of three complex exponentials:

$$w[n] = \frac{1}{2} - \frac{1}{4}U^n - \frac{1}{4}U^{-n}$$

where as before, U is the unit-magnitude complex number with argument $2\pi/N$. We can now calculate the windowed Fourier transform of a sinusoid Z^n with angular frequency α as before. The phases come out messy and we'll replace them with simplified approximations:

$$\mathcal{FT}\left\{w[n]Z^n\right\}(k)$$

$$= \mathcal{FT}\left\{\frac{1}{2}Z^n - \frac{1}{4}(UZ)^n - \frac{1}{4}(U^{-1}Z)^n\right\}(k)$$

$$\approx \left[\cos(\Phi(k)) + i\sin(\Phi(k))\right]M(k - \frac{\alpha}{\omega})$$

where the (approximate) phase term is:

$$\Phi(k) = -\pi \cdot (k - \frac{\alpha}{\omega})$$

and the magnitude function is:

$$M(k) = \left[\frac{1}{2}D_N(k) + \frac{1}{4}D_N(k+1) + \frac{1}{4}D_N(k-1)\right]$$

The magnitude function $M(k)$ is graphed in Figure 9.5. The three Dirichlet kernel components are also shown separately.

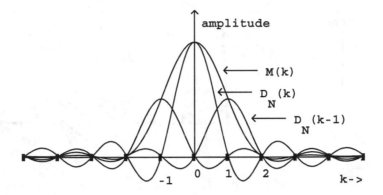

Figure 9.5: The magnitude M(k) of the Fourier transform of the Hann window function. It is the sum of three (shifted and magnified) copies of the Dirichlet kernel D_N, with $N = 100$.

The main lobe of $M(k)$ is four harmonics wide, twice the width of the main lobe of the Dirichlet kernel. The sidelobes, on the other hand, have much smaller magnitude. Each sidelobe of $M(k)$ is a sum of three sidelobes of $D_n(k)$, one attenuated by 1/2 and the others, opposite in sign, attenuated by 1/4. They do not cancel out perfectly but they do cancel out fairly well.

The sidelobes reach their maximum amplitudes near their midpoints, and we can estimate their amplitudes there, using the approximation:

$$D_N(k) \approx \frac{N \sin(\pi k)}{\pi k}$$

Setting $k = 3/2, 5/2, \ldots$ gives sidelobe amplitudes, relative to the peak height N, of:

$$\frac{2}{3\pi} \approx -13\text{dB}, \quad \frac{2}{5\pi} \approx -18\text{dB}, \quad \frac{2}{7\pi} \approx -21\text{dB}, \quad \frac{2}{9\pi} \approx -23\text{dB}, \ldots$$

The sidelobes drop off progressively more slowly so that the tenth one is only attenuated about 30 dB and the 32nd one about -40 dB. On the other hand, the Hann window sidelobes are attenuated by:

$$\frac{2}{5\pi} - \frac{1}{2}[\frac{2}{3\pi} + \frac{2}{7\pi}] \approx -32.30\text{dB}$$

and $-42, -49, -54$, and -59 dB for the next four sidelobes.

This shows that applying a Hann window before taking the Fourier transform will better allow us to isolate sinusoidal components. If a signal

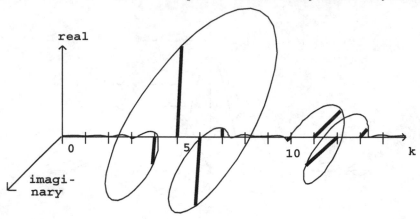

Figure 9.6: The Hann-windowed Fourier transform of a signal with two sinusoidal components, at frequencies 5.3 and 10.6 times the fundamental, and with different complex amplitudes.

has many sinusoidal components, the sidelobes engendered by each one will interfere with the main lobe of all the others. Reducing the amplitude of the sidelobes reduces this interference.

Figure 9.6 shows a Hann-windowed Fourier analysis of a signal with two sinusoidal components. The two are separated by about 5 times the fundamental frequency ω, and for each we see clearly the shape of the Hann window's Fourier transform. Four points of the Fourier analysis lie within the main lobe of $M(k)$ corresponding to each sinusoid. The amplitude and phase of the individual sinusoids are reflected in those of the (four-point-wide) peaks. The four points within a peak which happen to fall at integer values k are successively about one half cycle out of phase.

To fully resolve the partials of a signal, we should choose an analysis size N large enough so that $\omega = 2\pi/N$ is no more than a quarter of the frequency separation between neighboring partials. For a periodic signal, for example, the partials are separated by the fundamental frequency. For the analysis to fully resolve the partials, the analysis period N must be at least four periods of the signal.

In some applications it works to allow the peaks to overlap as long as the center of each peak is isolated from all the other peaks; in this case the four-period rule may be relaxed to three or even slightly less.

9.4 Fourier Analysis and Reconstruction of Audio Signals

Fourier analysis can sometimes be used to resolve the component sinusoids in an audio signal. Even when it can't go that far, it can separate a signal into frequency regions, in the sense that for each k, the kth point of the Fourier transform would be affected only by components close to the nominal frequency $k\omega$. This suggests many interesting operations we could perform on a signal by taking its Fourier transform, transforming the result, and then reconstructing a new, transformed, signal from the modified transform.

Figure 9.7 shows how to carry out a Fourier analysis, modification, and reconstruction of an audio signal. The first step is to divide the signal into *windows*, which are segments of the signal, of N samples each, usually with some overlap. Each window is then shaped by multiplying it by a windowing function (Hann, for example). Then the Fourier transform is calculated for the N points $k = 0, 1, \ldots, N - 1$. (Sometimes it is desirable to calculate the Fourier transform for more points than this, but these N points will suffice here.)

The Fourier analysis gives us a two-dimensional array of complex numbers. Let H denote the *hop size*, the number of samples each window is advanced past the previous window. Then for each $m = \ldots, 0, 1, \ldots$, the mth window consists of the N points starting at the point mH. The nth point of the mth window is $mH + n$. The windowed Fourier transform is thus equal to:

$$S[m, k] = \mathcal{FT}\{w(n)X[n - mH]\}(k)$$

This is both a function of time (m, in units of H samples) and of frequency (k, as a multiple of the fundamental frequency ω). Fixing the frame number m and looking at the windowed Fourier transform as a function of k:

$$S[k] = S[m, k]$$

gives us a measure of the momentary spectrum of the signal $X[n]$. On the other hand, fixing a frequency k we can look at it as the kth channel of an N-channel signal:

$$C[m] = S[m, k]$$

From this point of view, the windowed Fourier transform separates the original signal $X[n]$ into N narrow frequency regions, called *bands*.

Having computed the windowed Fourier transform, we next apply any desired modification. In the figure, the modification is simply to replace

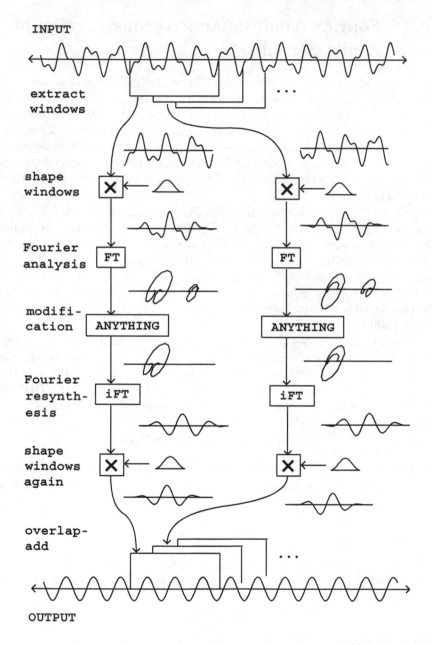

Figure 9.7: Sliding-window analysis and resynthesis of an audio signal using Fourier transforms. In this example the signal is filtered by multiplying the Fourier transform with a desired frequency response.

the upper half of the spectrum by zero, which gives a highly selective low-pass filter. (Two other possible modifications, narrow-band companding and vocoding, are described in the following sections.)

Finally we reconstruct an output signal. To do this we apply the inverse of the Fourier transform (labeled "iFT" in the figure). As shown in Section 9.1.2 this can be done by taking another Fourier transform, normalizing, and flipping the result backwards. In case the reconstructed window does not go smoothly to zero at its two ends, we apply the Hann windowing function a second time. Doing this to each successive window of the input, we then add the outputs, using the same overlap as for the analysis.

If we use the Hann window and an overlap of four (that is, choose N a multiple of four and space each window $H = N/4$ samples past the previous one), we can reconstruct the original signal faithfully by omitting the "modification" step. This is because the iFT undoes the work of the FT, and so we are multiplying each window by the Hann function squared. The output is thus the input, times the Hann window function squared, overlap-added by four. An easy check shows that this comes to the constant $3/2$, so the output equals the input times a constant factor.

The ability to reconstruct the input signal exactly is useful because some types of modification may be done by degrees, and so the output can be made to vary smoothly between the input and some transformed version of it.

9.4.1 Narrow-band companding

A *compander* is a tool that amplifies a signal with a variable gain, depending on the signal's measured amplitude. The term is a contraction of "compressor" and "expander". A compressor's gain decreases as the input level increases, so that the *dynamic range*, that is, the overall variation in signal level, is reduced. An expander does the reverse, increasing the dynamic range. Frequently the gain depends not only on the immediate signal level but on its history; for instance the rate of change might be limited or there might be a time delay.

By using Fourier analysis and resynthesis, we can do companding individually on narrow-band channels. If $C[m]$ is one such band, we apply a gain $g[m]$ to it, to give $g[m]C[m]$. Although $C[m]$ is a complex number, the gain is a non-negative real number. In general the gain could be a function not only of $C[m]$ but also of any or all the previous samples in the channel: $C[m-1]$, $C[m-2]$, and so on. Here we'll consider the simplest situation where the gain is simply a function of the magnitude of the current sample: $|C[m]|$.

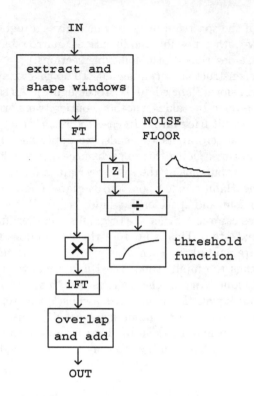

Figure 9.8: Block diagram for narrow-band noise suppression by companding.

The patch diagrammed in Figure 9.8 shows one very useful application of companding, called a *noise gate*. Here the gain $g[m]$ depends on the channel amplitude $C[m]$ and a noise floor which is a function f of the channel number k. For clarity we will apply the frequency subscript k to the gain, now written as $g[m, k]$, and to the windowed Fourier transform $S[m, k] = C[m]$. The gain is given by:

$$g[m, k] = \begin{cases} 1 - f[k]/|S[m, k]| & |S[m, k]| > f[k] \\ 0 & \text{otherwise} \end{cases}$$

Whenever the magnitude $S[m, k]$ is less than the threshold $f[k]$ the gain is zero and so the amplitude $S[m, k]$ is replaced by zero. Otherwise, multiplying the amplitude by $g[m, k]$ reduces the the magnitude downward to $|S[m, k]| - f[k]$. Since the gain is a non-negative real number, the phase is preserved.

In the figure, the gain is computed as a thresholding function of the ratio $x = |S[m, k]|/f[k]$ of the signal magnitude to the noise floor; the threshold is $g(x) = 1 - 1/x$ when $x < 1$ and zero otherwise, although other thresholding functions could easily be substituted.

This technique is useful for removing noise from a recorded sound. We either measure or guess values of the noise floor $f[k]$. Because of the design of the gain function $g[m, k]$, only amplitudes which are above the noise floor reach the output. Since this is done on narrow frequency bands, it is sometimes possible to remove most of the noise even while the signal itself, in the frequency ranges where it is louder than the noise floor, is mostly preserved.

The technique is also useful as preparation before applying a non-linear operation, such as distortion, to a sound. It is often best to distort only the most salient frequencies of the sound. Subtracting the noise-gated sound from the original then gives a residual signal which can be passed through undistorted.

9.4.2 Timbre stamping (classical vocoder)

A second application of Fourier analysis and resynthesis is a time-varying filter capable of making one sound take on the evolving spectral envelope of another. This is widely known in electronic music circles as a *vocoder*, named, not quite accurately, after the original Bell Laboratories vocal analysis/synthesis device. The technique described here is more accurately called *timbre stamping*. Two input signals are used, one to be filtered, and the other to control the filter via its time-varying spectral envelope. The windowed Fourier transform is used both on the control signal input to estimate its spectral envelope, and on the filter input in order to apply the filter.

A block diagram for timbre stamping is shown in Figure 9.9. As in the previous example, the timbre stamp acts by multiplying the complex-valued windowed Fourier transform of the filter input by non-negative real numbers, hence changing their magnitudes but leaving their phases intact. Here the twist is that we want simply to replace the magnitudes of the original, $|S[m, k]|$, with magnitudes obtained from the control input (call them $|T[m, k]|$, say). The necessary gain would thus be,

$$g[m, k] = \frac{|T[m, k]|}{|S[m, k]|}$$

In practice it is best to limit the gain to some maximum value (which might depend on frequency) since otherwise channels containing nothing

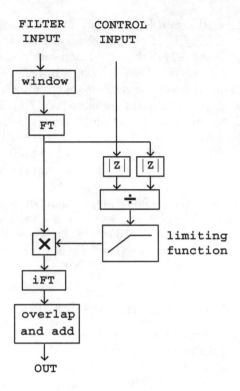

Figure 9.9: Block diagram for timbre stamping (AKA "vocoding").

but noise, sidelobes, or even truncation error might be raised to audibility. So a suitable limiting function is applied to the gain before using it.

9.5   Phase

So far we have operated on signals by altering the magnitudes of their windowed Fourier transforms, but leaving phases intact. The magnitudes encode the spectral envelope of the sound. The phases, on the other hand, encode frequency and time, in the sense that phase change from one window to a different one accumulates, over time, according to frequency. To make a transformation that allows independent control over frequency and time requires analyzing and reconstructing the phase.

In the analysis/synthesis examples of the previous section, the phase of the output is copied directly from the phase of the input. This is appro-

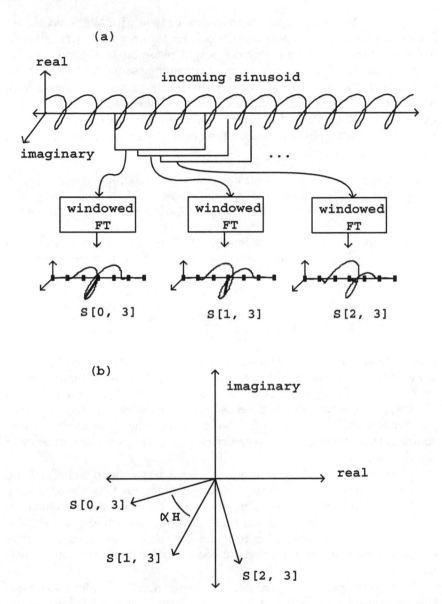

Figure 9.10: Phase in windowed Fourier analysis: (a) a complex sinusoid analyzed on three successive windows; (b) the result for a single channel (k=3), for the three windows.

priate when the output signal corresponds in time with the input signal. Sometimes time modifications are desired, for instance to do time stretching or contraction. Alternatively the output phase might depend on more than one input, for instance to morph between one sound and another.

Figure 9.10 shows how the phase of the Fourier transform changes from window to window, given a complex sinusoid as input. The sinusoid's frequency is $\alpha = 3\omega$, so that the peak in the Fourier transform is centered at $k = 3$. If the initial phase is ϕ, then the neighboring phases can be filled in as:

$$\angle S[0,2] = \phi + \pi \qquad \angle S[0,3] = \phi \qquad \angle S[0,4] = \phi + \pi$$
$$\angle S[1,2] = \phi + H\alpha + \pi \qquad \angle S[1,3] = \phi + H\alpha \qquad \angle S[1,4] = \phi + H\alpha + \pi$$
$$\angle S[2,2] = \phi + 2H\alpha + \pi \qquad \angle S[2,3] = \phi + 2H\alpha \qquad \angle S[2,4] = \phi + 2H\alpha + \pi$$

This gives an excellent way of estimating the frequency α: pick any channel whose amplitude is dominated by the sinusoid and subtract two successive phase to get $H\alpha$:

$$H\alpha = \angle S[1,3] - \angle S[0,3]$$

$$\alpha = \frac{\angle S[1,3] - \angle S[0,3] + 2p\pi}{H}$$

where p is an integer. There are H possible frequencies, spaced by $2\pi/H$. If we are using an overlap of 4, that is, $H = N/4$, the frequencies are spaced by $8\pi/N = 4\omega$. Happily, this is the width of the main lobe for the Hann window, so no more than one possible value of α can explain any measured phase difference within the main lobe of a peak. The correct value of p to choose is that which gives a frequency closest to the nominal frequency of the channel, $k\omega$.

When computing phases for synthesizing a new or modified signal, we want to maintain the appropriate phase relationships between successive resynthesis windows, and also, simultaneously, between adjacent channels. These two sets of relationships are not always compatible, however. We will make it our first obligation to honor the relations between successive resynthesis windows, and worry about phase relationships between channels afterward.

Suppose we want to construct the mth spectrum $S[m, k]$ for resynthesis (having already constructed the previous one, number $m - 1$). Suppose we wish the phase relationships between windows $m - 1$ and m to be those of a signal $x[n]$, but that the phases of window number $m - 1$ might have come from somewhere else and can't be assumed to be in line with our wishes.

To find out how much the phase of each channel should differ from the previous one, we do two analyses of the signal $x[n]$, separated by the same

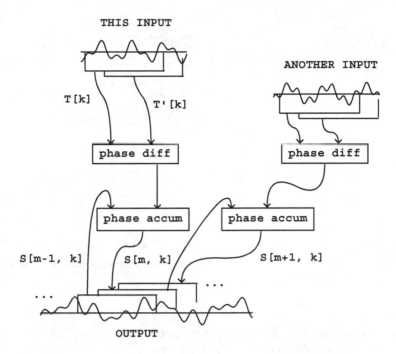

Figure 9.11: Propagating phases in resynthesis. Each phase, such as that of $S[m, k]$ here, depends on the previous output phase and the difference of the input phases.

hop size H that we're using for resynthesis:

$$T[k] = \mathcal{FT}(W(n)X[n])(k)$$

$$T'[k] = \mathcal{FT}(W(n)X[n+H])(k)$$

Figure 9.11 shows the process of phase accumulation, in which the output phases each depend on the previous output phase and the phase difference for two windowed analyses of the input. Figure 9.12 illustrates the phase relationship in the complex plane. The phase of the new output $S[m, k]$ should be that of the previous one plus the difference between the phases of the two analyses:

$$\angle S[m, k] = \angle S[m - 1, k] + (\angle T'[k] - \angle T[k])$$

$$= \angle \left(\frac{S[m - 1, k]T'[k]}{T[k]} \right)$$

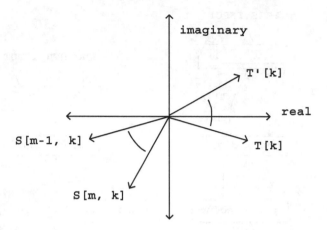

Figure 9.12: Phases of one channel of the analysis windows and two successive resynthesis windows.

Here we used the fact that multiplying or dividing two complex numbers gives the sum or difference of their arguments.

If the desired magnitude is a real number a, then we should set $S[m,k]$ to:

$$S[m,k] \;=\; a \cdot \left| \frac{S[m-1,k]T'[k]}{T[k]} \right|^{-1} \cdot \frac{S[m-1,k]T'[k]}{T[k]}$$

The magnitudes of the second and third terms cancel out, so that the magnitude of $S[m,k]$ reduces to a; the first two terms are real numbers so the argument is controlled by the last term.

If we want to end up with the magnitude from the spectrum T as well, we can set $a = |T'[k]|$ and simplify:

$$S[m,k] \;=\; \left| \frac{S[m-1,k]}{T[k]} \right|^{-1} \cdot \frac{S[m-1,k]T'[k]}{T[k]}$$

9.5.1   Phase relationships between channels

In the scheme above, the phase of each $S[m,k]$ depends only on the previous value for the same channel. The phase relationships between neighboring channels are left to chance. This sometimes works fine, but sometimes the incoherence of neighboring channels gives rise to an unintended chorus effect. We would ideally wish for $S[m,k]$ and $S[m,k+1]$ to have the same phase relationship as for $T'[k]$ and $T'[k+1]$, but also for the phase relationship between $S[m, k]$ and $S[m-1, k]$ to be the same as between $T'[k]$ and $T[k]$.

These $2N$ equations for N phases in general will have no solution, but we can alter the equation for $S[m, k]$ above so that whenever there happens to be a solution to the over-constrained system of equations, the reconstruction algorithm homes in on the solution. This approach is called *phase locking* [Puc95b], and has the virtue of simplicity although more sophisticated techniques are available [DL97]).

The desired output phase relation, at the frame $m-1$, is:

$$\angle T[k+1] - \angle T[k] = \angle S[m-1, k+1] - \angle S[m-1, k]$$

or, rearranging:

$$\angle \left\{ \frac{S[m-1, k+1]}{T[k+1]} \right\} = \angle \left\{ \frac{S[m-1, k]}{T[k]} \right\}$$

In other words, the phase of the quotient S/T should not depend on k. With this in mind, we can rewrite the recursion formula for $S[m, k]$:

$$S[m, k] = |R[k]|^{-1} \cdot R[k] T'[k]$$

with

$$R[k] = \frac{\overline{T[k]} \cdot S[m-1, k]}{|S[m-1, k]|}$$

and because of the previous equation, the $R[k]$ should all be in phase. The trick is now to replace $R[k]$ for each k with the sum of three neighboring ones. The computation is then:

$$S[m, k] = |R'[k]|^{-1} \cdot R'[k] T'[k]$$

with

$$R'[k] = R[k+1] + R[k] + R[k-1]$$

If the channels are already in the correct phase relationship, this has no effect (the resulting phase will be the same as if only $R[k]$ were used.) But in general the sum will share two terms in common with its neighbor at $k+1$:

$$R'[k+1] = R[k+2] + R[k+1] + R[k]$$

so that the R' will tend to point more in the same direction than the R do. Applying this iteratively will eventually line all the R' up to the same phase, as long as the phase relationships between the measured spectra T and T' allow it.

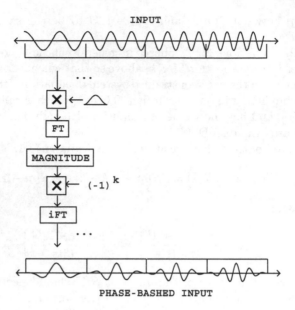

Figure 9.13: Phase-bashing a recorded sound (here, a sinusoid with rising frequency) to give a series of oscillator wavetables.

9.6   Phase Bashing

In Section 2.3 on enveloped sampling we saw how to make a periodic waveform from a recorded sound, thereby borrowing the timbre of the original sound but playing it at a specified pitch. If the window into the recorded sound is made to precess in time, the resulting timbre varies in imitation of the recorded sound.

One important problem arises, which is that if we take waveforms from different windows of a sample (or from different samples), there is no guarantee that the phases of the two match up. If they don't the result can be ugly, since the random phase changes are heard as frequency fluctuations. This can be corrected using Fourier analysis and resynthesis [Puc05].

Figure 9.13 shows a simple way to use Fourier analysis to align phases in a series of windows in a recording. We simply take the FFT of the window and then set each phase to zero for even values of k and π for odd ones. The phase at the center of the window is thus zero for both even and odd values of k. To set the phases (the arguments of the complex amplitudes in the spectrum) in the desired way, first we find the magnitude, which can be considered a complex number with argument zero. Then multiplying by

$(-1)^k$ adjusts the amplitude so that it is positive and negative in alternation. Then we take the inverse Fourier transform, without even bothering to window again on the way back; we will probably want to apply a windowing envelope later anyway as was shown in Figure 2.7. The results can be combined with the modulation techniques of Chapter 6 to yield powerful tools for vocal and other imitative synthesis.

9.7 Examples

Fourier analysis and resynthesis in Pd

Example I01.Fourier.analysis.pd (Figure 9.14, part a) demonstrates computing the Fourier transform of an audio signal using the `fft~` object:

`fft~`: Fast Fourier transform. The two inlets take audio signals representing the real and imaginary parts of a complex-valued signal. The window size N is given by Pd's block size. One Fourier transform is done on each block.

The Fast Fourier transform [SI03] reduces the computational cost of Fourier analysis in Pd to only that of between 5 and 15 `osc~` objects in typical configurations. The FFT algorithm in its simplest form takes N to be a power of two, which is also (normally) a constraint on block sizes in Pd.

Example I02.Hann.window.pd (Figure 9.14, parts b and c) shows how to control the block size using a `block~` object, how to apply a Hann window, and a different version of the Fourier transform. Part (b) shows the invocation of a subwindow which in turn is shown in part (c). New objects are:

`rfft~`: real Fast Fourier transform. The imaginary part of the input is assumed to be zero. Only the first $N/2 + 1$ channels of output are filled in (the others are determined by symmetry). This takes half the computation time of the (more general) `fft~`object.

`tabreceive~`: repeatedly outputs the contents of a wavetable. Each block of computation outputs the same first N samples of the table.

In this example, the table "$0-hann" holds a Hann window function of length 512, in agreement with the specified block size. The signal to be analyzed appears (from the parent patch) via the `inlet~` object. The channel amplitudes (the output of the `rfft~` object) are reduced to real-valued magnitudes: the real and imaginary parts are squared separately, the two squares are added, and the result passed to the `sqrt~` object. Finally the magnitude is written (controlled by a connection not shown in the figure) via `tabwrite~` to another table, "$0-magnitude", for graphing.

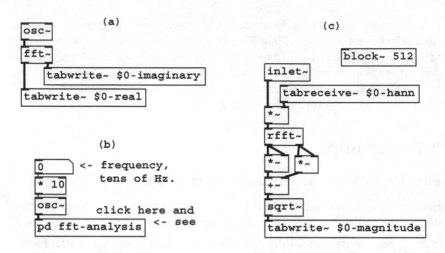

Figure 9.14: Fourier analysis in Pd: (a) the `fft~` object; (b) using a sub-window to control block size of the Fourier transform; (c) the subwindow, using a real Fourier transform (the `rfft~` object) and the Hann windowing function.

Example I03.resynthesis.pd (Figure 9.15) shows how to analyze and resynthesize an audio signal following the strategy of Figure 9.7. As before there is a sub-window to do the work at a block size appropriate to the task; the figure shows only the sub-window. We need one new object for the inverse Fourier transform:

`rifft~` : real inverse Fast Fourier transform. Using the first $N/2 + 1$ points of its inputs (taken to be a real/imaginary pair), and assuming the appropriate values for the other channels by symmetry, reconstructs a real-valued output. No normalization is done, so that a `rfft~`/`rifft~` pair together result in a gain of N. The `ifft~` object is also available which computes an unnormalized inverse for the `fft~` object, reconstructing a complex-valued output.

The `block~` object, in the subwindow, is invoked with a second argument which specifies an overlap factor of 4. This dictates that the sub-window will run four times every $N = 512$ samples, at regular intervals of 128 samples. The `inlet~` object does the necessary buffering and rearranging of samples so that its output always gives the 512 latest samples of input in order. In the other direction, the `outlet~` object adds segments of its previous four inputs to carry out the overlap-add scheme shown in Figure 9.7.

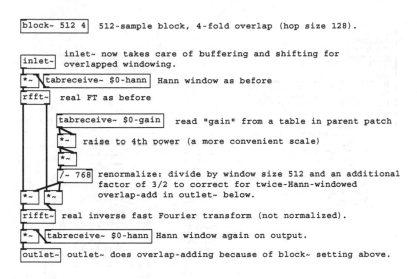

| block~ 512 4 | 512-sample block, 4-fold overlap (hop size 128). |

inlet~ now takes care of buffering and shifting for
| inlet~ | overlapped windowing.

| *~ \ tabreceive~ $0-hann | Hann window as before

| rfft~ | real FT as before

| tabreceive~ $0-gain | read "gain" from a table in parent patch

| *~ | raise to 4th power (a more convenient scale)

| *~ |

| /~ 768 | renormalize: divide by window size 512 and an additional
factor of 3/2 to correct for twice-Hann-windowed
overlap-add in outlet~ below.

| *~ | *~ |

| rifft~ | real inverse fast Fourier transform (not normalized).

| *~ \ tabreceive~ $0-hann | Hann window again on output.

| outlet~ | outlet~ does overlap-adding because of block~ setting above.

Figure 9.15: Fourier analysis and resynthesis, using block~ to specify an
overlap of 4, and rifft~ to reconstruct the signal after modification.

The 512-sample blocks are multiplied by the Hann window both at the
input and the output. If the rfft~ and rifft~ objects were connected
without any modifications in between, the output would faithfully reconstruct the input.

A modification is applied, however: each channel is multiplied by a
(positive real-valued) gain. The complex-valued amplitude for each channel
is scaled by separately multiplying the real and imaginary parts by the gain.
The gain (which depends on the channel) comes from another table, named
"$0-gain". The result is a graphical equalization filter; by mousing in the
graphical window for this table, you can design gain-frequency curves.

There is an inherent delay introduced by using block~ to increase the
block size (but none if it is used, as shown in Chapter 7, to reduce block
size relative to the parent window.) The delay can be measured from the
inlet to the outlet of the sub-patch, and is equal to the difference of the
two block sizes. In this example the buffering delay is 512-64=448 samples.
Blocking delay does not depend on overlap, only on block sizes.

Narrow-band companding: noise suppression

Example I04.noisegate.pd (Figure 9.16) shows an example of narrow-band
companding using Fourier analysis/resynthesis. (This is a realization of

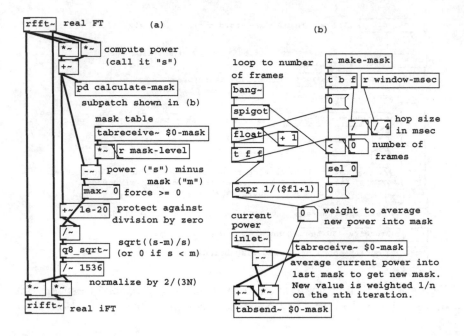

Figure 9.16: Noise suppression as an example of narrow-band companding: (a) analysis and reconstruction of the signal; (b) computation of the "mask".

the block diagram of Figure 9.8.) Part (a) of the figure shows a filter configuration similar to the previous example, except that the gain for each channel is now a function of the channel magnitude.

For each k, if we let $s[k]$ denote the power in channel k, and let $m[k]$ be a mask level (a level presumably somewhat higher than the noise power for channel k), then the gain in channel k is given by

$$\left\{ \begin{array}{ll} \sqrt{\frac{s[k]-m[k]}{s[k]}} & s[k] > m[k] \\ 0 & \text{otherwise} \end{array} \right.$$

The power in the kth channel is thus reduced by $m[k]$ if possible, and otherwise replaced by zero.

The mask itself is the product of the measured average noise in each channel, which is contained in the table "$0-mask", multiplied by a value named "mask-level". The average noise is measured in a subpatch (pd calculate-mask), whose contents are shown in part (b) of the figure. To compute the mask we are using two new new objects:

bang~ : send a bang in advance of each block of computation. The bang appears at the logical time of the first sample in each block (the earliest logical time whose control computation affects that block and not the previous one), following the scheme shown in Figure 3.2.

tabsend~ : the companion object for tabreceive~, repeatedly copies its input to the contents of a table, affecting up to the first N samples of the table.

The power averaging process is begun by sending a time duration in milliseconds to "make-mask". The patch computes the equivalent number of blocks b and generates a sequence of weights: $1, 1/2, 1/3, \ldots, 1/b$, by which each of the b following blocks' power is averaged into whatever the mask table held at the previous block. At the end of b blocks the table holds the equally-weighted average of all b power measurements. Thereafter, the weight for averaging new power measurements is zero, so the measured average stops evolving.

To use this patch for classical noise suppression requires at least a few seconds of recorded noise without the "signal" present. This is played into the patch, and its duration sent to "make-mask", so that the "$0-mask" table holds the average measured noise power for each channel. Then, making the assumption that the noisy part of the signal rarely exceeds 10 times its average power (for example), "mask-level" is set to 10, and the signal to be noise-suppressed is sent through part (a) of the patch. The noise will be almost all gone, but those channels in which the signal exceeds 20 times the noise power will only be attenuated by 3dB, and higher-power channels progressively less. (Of course, actual noise suppression might not be the most interesting application of the patch; one could try masking any signal from any other one.)

Timbre stamp ("vocoder")

Example I05.compressor.pd (Figure 9.17) is another channel compander which is presented in preparation for Example I06.timbre.stamp.pd, which we will examine next. This is a realization of the timbre stamp of Figure 9.9, slightly modified.

There are two inputs, one at left to be filtered (and whose Fourier transform is used for resynthesis after modifying the magnitudes), and one at right which acts as a control source. Roughly speaking, if the two magnitudes are $f[k]$ for the filter input and $c[k]$ for the control source, we just "whiten" the filter input, multiplying by $1/f[k]$, and then stamp the control magnitudes onto the result by further multiplying by $c[k]$. In practice, we must limit the gain to some reasonable maximum value. In this patch

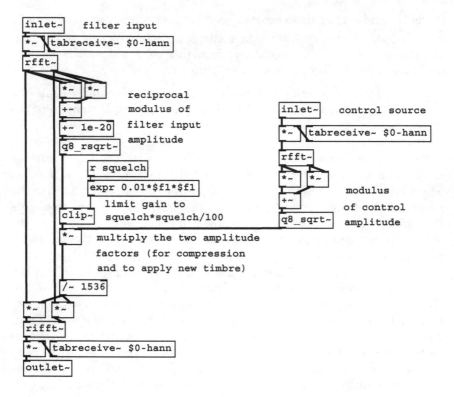

Figure 9.17: Timbre stamp.

this is done by limiting the whitening factor $1/f[k]$ to a specified maximum value using the clip~ object. The limit is controlled by the "squelch" parameter, which is squared and divided by 100 to map values from 0 to 100 to a useful range.

Another possible scheme is to limit the gain after forming the quotient $c[k]/f[k]$. The gain limitation may in either case be frequency dependent. It is also sometimes useful to raise the gain to a power p between 0 and 1; if 1, this is a timbre stamp and if 0, it passes the filter input through unchanged, and values in between give a smooth interpolation between the two.

Phase vocoder time bender

The phase vocoder usually refers to the general technique of passing from (complex-valued) channel amplitudes to pairs consisting of (real-valued) magnitudes and phase precession rates ("frequencies"), and back, as described in Figure 9.11 (Section 9.5). In Example I07.phase.vocoder.pd (Figure 9.18), we use this technique with the specific aim of time-stretching and/or time-contracting a recorded sound under real-time control. That is, we control, at any moment in real time, the location in the recorded sound we hear. Two new objects are used:

`lrshift~` : shift a block left or right (according to its creation argument). If the argument is positive, each block of the output is the input shifted that number of spaces to the right, filling zeros in as needed on the left. A negative argument shifts to the left, filling zeros in at the right.

`q8_rsqrt~` : quick and approximate reciprocal square root. Outputs the reciprocal of the square root of its input, good to about a part in 256, using much less computation than a full-precision square root and reciprocal would.

The process starts with a sub-patch, pd read-windows, that outputs two Hann-windowed blocks of the recorded sound, a "back" one and a "front" one 1/4 window further forward in the recording. The window shown uses the two outputs of the sub-patch to guide the amplitude and phase change of each channel of its own output.

The top two `tabreceive~` objects recall the previous block of complex amplitudes sent to the `rifft~` object at bottom, corresponding to $S[m - 1, k]$ in the discussion of Section 9.5. The patch as a whole computes $S[m, k]$ and then its Hann windowed inverse FT for output.

After normalizing $S[m - 1, k]$, its complex conjugate (the normalized inverse) is multiplied by the windowed Fourier transform of the "back" window $T[k]$, giving the product $R[k]$ of Page 288. Next, depending on the value of the parameter "lock", the computed value of $R[k]$ is conditionally replaced with the phase-locking version $R'[k]$. This is done using `lrshift~` objects, whose outputs are added into $R[k]$ if "lock" is set to one, or otherwise not if it is zero. The result is then normalized and multiplied by the Hann-windowed Fourier transform of the "front" window $(T'[k])$ to give $S[m, k]$.

Three other applications of Fourier analysis/resynthesis, not pictured here, are provided in the Pd examples. First, Example I08.pvoc.reverb.pd shows how to make a phase vocoder whose output recirculates as in a reverberator, except that individual channels are replaced by the input when it is more powerful than what is already recirculating. The result is a more coherent-sounding reverberation effect than can be made in the classical way using delay lines.

Example I09.sheep.from.goats.pd demonstrates the (imperfect) technique of separating pitched signals from noisy ones, channel by channel,

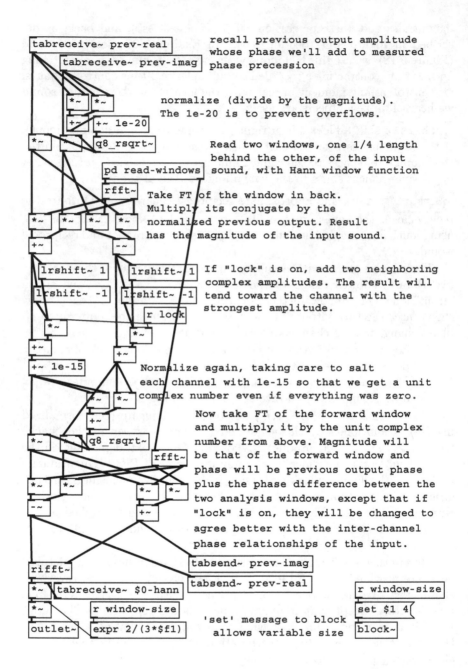

Figure 9.18: Phase vocoder for time stretching and contraction.

based on the phase coherence we should expect from a Hann-windowed sinusoid. If three adjacent channels are approximately π radians out of phase from each other, they are judged to belong to a sinusoidal peak. Channels belonging to sinusoidal peaks are replaced with zero to extract the noisy portion of the signal, or all others are replaced with zero to give the sinusoidal portion.

Example I10.phase.bash.pd returns to the wavetable looping sampler of Figure 2.7, and shows how to align the phases of the sample so that all components of the signal have zero phase at points 0, N, $2N$, and so on. In this way, two copies of a looping sampler placed N samples apart can be coherently cross-faded. A synthetic, pitched version of the original soundfile can be made using daisy-chained cross-fades.

Exercises

1. A signal $x[n]$ is 1 for $n = 0$ and 0 otherwise (an impulse). What is its (N-point) Fourier transform as a function of k?

2. Assuming further that N is an even number, what does the Fourier transform become if $x[n]$ is 1 at $n = N/2$ instead of at $n = 0$?

3. For what integer values of k is the Fourier transform of the N-point Hann window function nonzero?

4. In order to Fourier analyze a 100-Hertz periodic tone (at a sample rate of 44100 Hertz), using a Hann window, what value of N would be needed to completely resolve all the partials of the tone (in the sense of having non-overlapping peaks in the spectrum)?

5. Suppose an N-point Fourier transform is done on a complex sinusoid of frequency 2.5ω where $\omega = 2\pi/N$ is the fundamental frequency. What percentage of the signal energy lands in the main lobe, channels $k = 2$ and $k = 3$? If the signal is Hann windowed, what percentage of the energy is now in the main lobe (which is then channels 1 through 4)?

Chapter 10

Classical Waveforms

Up until now we have primarily taken three approaches to synthesizing repetitive waveforms: additive synthesis (Chapter 1), wavetable synthesis (Chapter 2), and waveshaping (Chapters 5 and 6). This chapter introduces a fourth approach, in which waveforms are built up explicitly from line segments with controllable endpoints. This approach is historically at least as important as the others, and was dominant during the analog synthesizer period, approximately 1965-1985. For lack of a better name, we'll use the term *classical waveforms* to denote waveforms composed of line segments.

They include the *sawtooth*, *triangle*, and *rectangle* waves pictured in Figure 10.1, among many other possibilities. The salient features of classical waveforms are either discontinuous jumps (changes in value) or corners (changes in slope). In the figure, the sawtooth and rectangle waves have jumps (once per cycle for the sawtooth, and twice for the rectangle), and constant slope elsewhere (negative for the sawtooth wave, zero for the rectangle wave). The triangle wave has no discontinuous jumps, but the slope changes discontinuously twice per cycle.

To use classical waveforms effectively, it is useful to understand how the shape of the waveform is reflected in its Fourier series. (To compute these we need background from Chapter 9, which is why this chapter appears here and not earlier.) We will also need strategies for digitally synthesizing classical waveforms. These waveforms prove to be much more susceptible to foldover problems than any we have treated before, so we will have to pay especially close attention to its control.

In general, our strategy for predicting and controlling foldover will be to consider first those sampled waveforms whose period is an integer N. Then if we want to obtain a waveform of a non-integral period (call it τ, say) we approximate τ as a quotient N/R of two integers. Conceptually

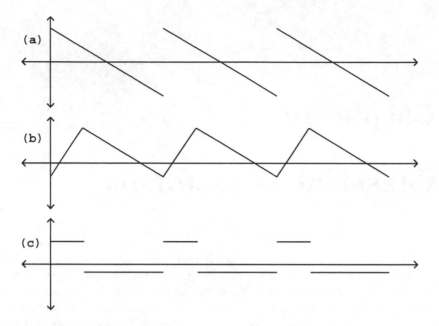

Figure 10.1: Classical waveforms: (a) the sawtooth, (b) the triangle, and (c) the rectangle wave, shown as functions of a continuous variable (not sampled).

at least, we can then synthesize the desired waveform with period N, and then take only one of each R samples of output. This last, down-sampling step is where the foldover is produced, and careful handling will help us control it.

10.1   Symmetries and Fourier Series

Before making a quantitative analysis of the Fourier series of the classical waveforms, we pause to make two useful observations about symmetries in waveforms and the corresponding symmetries in the Fourier series. First, a Fourier series might consist only of even or odd-numbered harmonics; this is reflected in symmetries comparing a waveform to its displacement by half a cycle. Second, the Fourier series may contain only real-valued or pure imaginary-valued coefficients (corresponding to the cosine or sine functions). This is reflected in symmetries comparing the waveform to its reversal in time.

In this section we will assume that our waveform has an integer period N, and furthermore, for simplicity, that N is even (if it isn't we can just up-sample by a factor of two). We know from Chapter 9 that any (real or complex valued) waveform $X[n]$ can be written as a Fourier series (whose coefficients we'll denote by $A[k]$):

$$X[n] = A[0] + A[1]U^n + \cdots + A[N-1]U^{(N-1)n}$$

or, equivalently,

$$X[n] = A[0] + A[1](\cos(\omega n) + i\sin(\omega n)) + \cdots$$

$$+ A[N-1](\cos(\omega(N-1)n) + i\sin(\omega(N-1)n))$$

where $\omega = 2\pi/N$ is the fundamental frequency of the waveform, and

$$U = \cos(\omega) + i\sin(\omega)$$

is the unit-magnitude complex number whose argument is ω.

To analyze the first symmetry we delay the signal $X[n]$ by a half-cycle. Since $U^{N/2} = -1$ we get:

$$X[n + N/2] = A[0] - A[1]U^n + A[2]U^{2n} \pm \cdots$$

$$+ A[N-2]U^{(N-2)n} - A[N-1]U^{(N-1)n}$$

In effect, a half-cycle delay changes the sign of every other term in the Fourier series. We combine this with the original series in two different ways. Letting X' denote half the sum of the two:

$$X'[n] = \frac{X[n] + X[n+N/2]}{2} = A[0] + A[2]U^{2n} + \cdots + A[N-2]U^{(N-2)n}$$

and X'' half the difference:

$$X''[n] = \frac{X[n] - X[n+N/2]}{2} = A[1]U^n + A[3]U^{3n} + \cdots + A[N-1]U^{(N-1)n}$$

we see that X' consists only of even-numbered harmonics (including DC) and X'' only of odd ones.

Furthermore, if X happens to be equal to itself shifted a half cycle, that is, if $X[n] = X[n+N/2]$, then (looking at the definitions of X' and X'') we get $X'[n] = X[n]$ and $X''[n] = 0$. This implies that, in this case, $X[n]$ has only even numbered harmonics. Indeed, this should be no surprise, since in this case $X[n]$ would have to repeat every $N/2$ samples, so its fundamental frequency is twice as high as normal for period N.

In the same way, if $X[n] = -X[n + N/2]$, then X can have only odd-numbered harmonics. This allows us easily to split any desired waveform into its even- and odd-numbered harmonics. (This is equivalent to using a comb filter to extract even or odd harmonics; see Chapter 7.)

To derive the second symmetry relation we compare $X[n]$ with its time reversal, $X[-n]$ (or, equivalently, since X repeats every N samples, with $X[N - n]$). The Fourier series becomes:

$$X[-n] = A[0] + A[1](\cos(\omega n) - i\sin(\omega n)) + \cdots$$

$$+A[N - 1](\cos(\omega(N - 1)n) - i\sin(\omega(N - 1)n))$$

(since the cosine function is even and the sine function is odd). In the same way as before we can extract the cosines by forming $X'[n]$ as half the sum:

$$X'[n] = \frac{X[n] + X[-n]}{2} = A[0] + A[1]\cos(\omega n) + \cdots + A[N-1]\cos(\omega(N-1)n)$$

and $X''[n]$ as half the difference divided by i:

$$X''[n] = \frac{X[n] - X[-n]}{2i} = A[1]\sin(\omega n) + \cdots + A[N - 1]\sin(\omega(N - 1)n)$$

So if $X[n]$ satisfies $X[-n] = X[n]$ the Fourier series consists of cosine terms only; if $X[-n] = -X[n]$ it consists of sine terms only; and as before we can decompose any $X[n]$ (that repeats every N samples) as a sum of the two.

10.1.1   Sawtooth waves and symmetry

As an example, we apply the shift symmetry (even and odd harmonics) to a sawtooth wave. Figure 10.2 (part a) shows the original sawtooth wave and part (b) shows the result of shifting by a half cycle. The sum of the two (part c) drops discontinuously whenever either one of the two copies does so, and traces a line segment whenever both component sawtooth waves do; so it in turn becomes a sawtooth wave, of half the original period (twice the fundamental frequency). Subtracting the two sawtooth waves (part d) gives a waveform with slope zero except at the discontinuities. The discontinuities coming from the original sawtooth wave jump in the same direction (negative to positive), but those coming from the shifted one are negated and jump from positive to negative. The result is a *square wave*, a particular rectangle wave in which the two component segments have the same duration.

This symmetry was used to great effect in the design of Buchla analog synthesizers; instead of offering a single sawtooth generator, Buchla

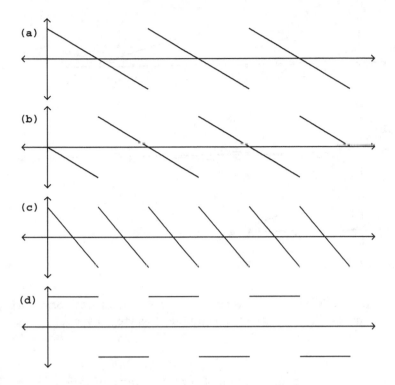

Figure 10.2: Using a symmetry relation to extract even and odd harmonics from a sawtooth wave: (a) the original sawtooth wave; (b) shifted by 1/2 cycle; (c) their sum (another sawtooth wave at twice the frequency); (d) their difference (a square wave).

designed an oscillator that outputs the even and odd harmonic portions separately, so that cross-fading between the two allows a continuous control over the relative strengths of the even and odd harmonics in the analog waveform.

10.2 Dissecting Classical Waveforms

Among the several conclusions we can draw from the even/odd harmonic decomposition of the sawtooth wave (Figure 10.2), one is that a square wave can be decomposed into a linear combination of two sawtooth waves. We can carry this idea further, and show how to compose any classical waveform having only jumps (discontinuities in value) but no corners

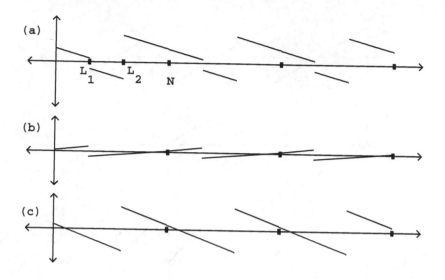

Figure 10.3: Dissecting a waveform: (a) the original waveform with two discontinuities; (b and c) the two component sawtooth waves.

(discontinuities in slope) as a sum of sawtooth waves of various phases and amplitudes. We then develop the idea further, showing how to generate waveforms with corners (either in addition to, or instead of, jumps) using another elementary waveform we'll call the *parabolic wave.*

Suppose first that a waveform of period N has discontinuities at j different points, $L_1, \ldots, L_j$, all lying on the cycle between 0 and N, at which the waveform jumps by values $d_1, \ldots, d_j$. A negative value of d_1, for instance, would mean that the waveform jumps from a higher to a lower value at the point L_1, and a positive value of d_1 would mean a jump from a lower to a higher value.

For instance, Figure 10.3 (part a) shows a classical waveform with two jumps: $(L_1, d_1) = (0.3N, -0.3)$ and $(L_2, d_2) = (0.6N, 1.3)$. Parts (b) and (c) show sawtooth waves, each with one of the two jumps. The sum of the two sawtooth waves reconstructs the waveform of part (a), except for a possible constant (DC) offset.

The sawtooth wave with a jump of one unit at the point zero is given by

$$s[n] = n/N - 1/2$$

over the period $0 \le n \le N - 1$, and repeats for other values of n. A

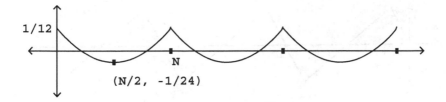

Figure 10.4: The parabolic wave.

sawtooth wave with a jump (L, d) is given by $s'[n] = ds[n - L]$. The sum of all the component sawtooth waves is:

$$x[n] = d_1 s[n - L_1] + \cdots + d_j s[n - L_j]$$

The slopes of the segments of the waveform of part (a) of the figure are all the same, equal to the sum of the slopes of the component sawtooth waves:

$$-\frac{d_1 + \cdots + d_j}{N}$$

Square and rectangle waves have horizontal line segments (slope zero); for this to happen in general the jumps must add to zero: $d_1 + \cdots + d_j = 0$.

To decompose classical waveforms with corners we use the parabolic wave, which, over a single period from 0 to N, is equal to

$$p[n] = \frac{1}{2}\left(\frac{n}{N} - \frac{1}{2}\right)^2 - \frac{1}{24}$$

as shown in Figure 10.4. It is a second-degree (quadratic) polynomial in the variable n, arranged so that it reaches a maximum halfway through the cycle at $n = N/2$, the DC component is zero (or in other words, the average value over one cycle of the waveform is zero), and so that the slope changes discontinuously by $-1/N$ at the beginning of the cycle.

To construct a waveform with any desired number of corners (suppose they are at the points $M_i, \ldots, M_l$, with slope changes equal to $c_1, \ldots, c_l$), we sum up the necessary parabolic waves:

$$x[n] = -N c_1 p[n - M_1] - \cdots - N c_l p[n - M_l]$$

An example is shown graphically in Figure 10.5.

If the sum $x[n]$ is to contain line segments (not segments of curves), the n^2 terms in the sum must sum to zero. From the expansion of $x[n]$ above, this implies that $c_1 + \cdots + c_l = 0$. Sums obtained from existing classical

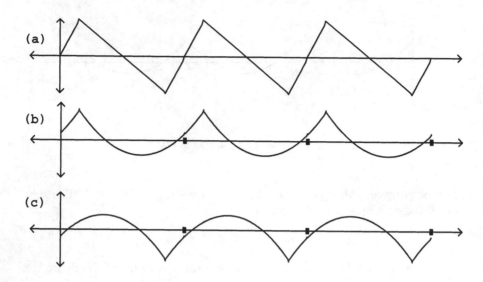

Figure 10.5: Decomposing a triangle wave (part a) into two parabolic waves (b and c).

waveforms (as in the figure) will always satisfy this condition because the changes in slope, over a cycle, must all add to zero for the waveform to connect with itself.

10.3 Fourier Series of the Elementary Waveforms

In general, given a repeating waveform $X[n]$, we can evaluate its Fourier series coefficients $A[k]$ by directly evaluating the Fourier transform:

$$A[k] = \frac{1}{N}\mathcal{FT}\{X[n]\}(k)$$

$$= \frac{1}{N}\left[X[0] + U^{-k}X[1] + \cdots + U^{-(N-1)k}X[N-1]\right]$$

but doing this directly for sawtooth and parabolic waves will require pages of algebra (somewhat less if we were willing resort to differential calculus). Instead, we rely on properties of the Fourier transform to relate the transform of a signal $x[n]$ with its *first difference*, defined as $x[n] - x[n-1]$. The first difference of the parabolic wave will turn out to be a sawtooth, and that of a sawtooth will be simple enough to evaluate directly, and thus we'll get the desired Fourier series.

In general, to evaluate the strength of the kth harmonic, we'll make the assumption that N is much larger than k, or equivalently, that k/N is negligible.

We start from the Time Shift Formula for Fourier Transforms (Page 272) setting the time shift to one sample:

$$\mathcal{FT}\{x[n-1]\} = [\cos(k\omega) - i\sin(k\omega)]\,\mathcal{FT}\{x[n]\}$$

$$\approx (1 - i\omega k)\mathcal{FT}\{x[n]\}$$

Here we're using the assumption that, because N is much larger than k, $k\omega = 2\pi k/N$ is much smaller than unity and we can make approximations:

$$\cos(k\omega) \approx 1 \ , \ \sin(k\omega) \approx k\omega$$

which are good to within a small error, on the order of $(k/N)^2$. Now we plug this result in to evaluate:

$$\mathcal{FT}\{x[n] - x[n-1]\} \approx i\omega k\mathcal{FT}\{x[n]\}$$

10.3.1 Sawtooth wave

First we apply this to the sawtooth wave $s[n]$. For $0 \le n < N$ we have:

$$s[n] - s[n-1] = -\frac{1}{N} + \begin{cases} 1 & n = 0 \\ 0 & \text{otherwise} \end{cases}$$

Ignoring the constant offset of $-\frac{1}{N}$, this gives an *impulse*, zero everywhere except one sample per cycle. The summation in the Fourier transform only has one term, and we get:

$$\mathcal{FT}\{s[n] - s[n-1]\}(k) = 1, \ k \neq 0, \ -N < k < N$$

We then apply the difference formula backward to get:

$$\mathcal{FT}\{s[n]\}(k) \approx \frac{1}{i\omega k} = \frac{-iN}{2\pi k}$$

valid for integer values of k, small compared to N, but with $k \neq 0$. (To get the second form of the expression we plugged in $\omega = 2\pi/N$ and $1/i = -i$.)

This analysis doesn't give us the DC component $\mathcal{FT}\{s[n]\}(0)$, because we would have had to divide by $k = 0$. Instead, we can evaluate the DC term directly as the sum of all the points of the waveform: it's approximately zero by symmetry.

To get a Fourier series in terms of familiar real-valued sine and cosine functions, we combine corresponding terms for negative and positive values of k. The first harmonic ($k = \pm 1$) is:

$$\frac{1}{N} \left[\mathcal{FT}\{s[n]\}(1) \cdot U^n + \mathcal{FT}\{s[n]\}(-1) \cdot U^{-n} \right]$$

$$\approx \frac{-i}{2\pi} \left[U^n - U^{-n} \right]$$

$$= \frac{\sin(\omega n)}{\pi}$$

and similarly the kth harmonic is

$$\frac{\sin(k\omega n)}{k\pi}$$

so the entire Fourier series is:

$$s[n] \approx \frac{1}{\pi} \left[\sin(\omega n) + \frac{\sin(2\omega n)}{2} + \frac{\sin(3\omega n)}{3} + \cdots \right]$$

10.3.2   Parabolic wave

The same analysis, with some differences in sign and normalization, works for parabolic waves. First we compute the difference:

$$p[n] - p[n-1] = \frac{\left(\frac{n}{N} - \frac{1}{2}\right)^2 - \left(\frac{n-1}{N} - \frac{1}{2}\right)^2}{2}$$

$$= \frac{\left(\frac{n}{N} - \frac{N}{2N}\right)^2 - \left(\frac{n}{N} - \frac{N-2}{2N}\right)^2}{2}$$

$$= \frac{\frac{2n}{N^2} - \frac{1}{N} + \frac{1}{N^2}}{2}$$

$$\approx -s[n]/N.$$

So (again for $k \neq 0$, small compared to N) we get:

$$\mathcal{FT}\{p[n]\}(k) \approx \frac{-1}{N} \cdot \frac{-iN}{2\pi k} \cdot \mathcal{FT}\{s[n]\}(k)$$

$$\approx \frac{-1}{N} \cdot \frac{-iN}{2\pi k} \cdot \frac{-iN}{2\pi k}$$

$$= \frac{N}{4\pi^2 k^2}$$

and as before we get the Fourier series:

$$p[n] \approx \frac{1}{2\pi^2} \left[\cos(\omega n) + \frac{\cos(2\omega n)}{4} + \frac{\cos(3\omega n)}{9} + \cdots \right]$$

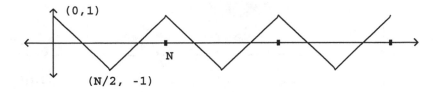

Figure 10.6: Symmetric triangle wave, obtained by superposing parabolic waves with (M, c) pairs equal to $(0, 8)$ and $(N/2, -8)$.

10.3.3 Square and symmetric triangle waves

To see how to obtain Fourier series for classical waveforms in general, consider first the square wave,

$$x[n] = s[n] - s[n - \frac{N}{2}]$$

equal to $1/2$ for the first half cycle ($0 <= n < N/2$) and $-1/2$ for the rest. We get the Fourier series by plugging in the Fourier series for $s[n]$ twice:

$$x[n] \approx \frac{1}{\pi} \left[\sin(\omega n) + \frac{\sin(2\omega n)}{2} + \frac{\sin(3\omega n)}{3} + \cdots \right.$$

$$\left. -\sin(\omega n) + \frac{\sin(2\omega n)}{2} - \frac{\sin(3\omega n)}{3} \pm \cdots \right]$$

$$= \frac{2}{\pi} \left[\sin(\omega n) + \frac{\sin(3\omega n)}{3} + \frac{\sin(5\omega n)}{5} + \cdots \right]$$

The symmetric triangle wave (Figure 10.6) given by

$$x[n] = 8p[n] - 8p[n - \frac{N}{2}]$$

similarly comes to

$$x[n] \approx \frac{8}{\pi^2} \left[\cos(\omega n) + \frac{\cos(3\omega n)}{9} + \frac{\cos(5\omega n)}{25} + \cdots \right]$$

10.3.4 General (non-symmetric) triangle wave

A general, non-symmetric triangle wave appears in Figure 10.7. Here we have arranged the cycle so that, first, the DC component is zero (so that

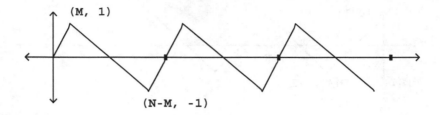

Figure 10.7: Non-symmetric triangle wave, with vertices at $(M, 1)$ and $(N - M, -1)$.

the two corners have equal and opposite heights), and second, so that the midpoint of the shorter segment goes through the point $(0, 0)$.

The two line segments have slopes equal to $1/M$ and $-2/(N - 2M)$, so the decomposition into component parabolic waves is given by:

$$x[n] = \frac{N^2}{MN - 2M^2}(p[n - M] - p[n + M])$$

(here we're using the periodicity of $p[n]$ to replace $p[n - (N - M)]$ by $p[n + M]$).)

The most general way of dealing with linear combinations of elementary (parabolic and/or sawtooth) waves is to go back to the complex Fourier series, as we did in finding the series for the elementary waves themselves. But in this particular case we can use a trigonometric identity to avoid the extra work of converting back and forth. First we plug in the real-valued Fourier series:

$$x[n] = \frac{N^2}{2\pi^2(MN - 2M^2)}\left[\cos(\omega(n - M)) - \cos(\omega(n + M))\right.$$
$$\left. + \frac{\cos(2\omega(n - M)) - \cos(2\omega(n + M))}{4} + \cdots\right]$$

Now we use the identity,

$$\cos(a) - \cos(b) = 2\sin(\frac{b - a}{2})\sin(\frac{a + b}{2})$$

so that, for example,

$$\cos(\omega(n - M)) - \cos(\omega(n + M)) = 2\sin(2\pi M/N)\sin(\omega n)$$

(Here again we used the definition of $\omega = 2\pi/N$.) This is a simplification since the first sine term does not depend on n; it's just an amplitude term.

Applying the identity to all the terms of the expansion for $x[n]$ gives:

$$x[n] = a[1]\sin(\omega n) + a[2]\sin(2\omega n) + \cdots$$

where the amplitudes of the components are given by:

$$a[k] = \frac{1}{\pi^2(M/N - 2(M/N)^2)} \cdot \frac{\sin(2\pi k M/N)}{k^2}$$

Notice that the result does not depend separately on the values of M and N, but only on their ratio, M/N (this is not surprising because the shape of the waveform depends on this ratio). If we look at small values of k:

$$k < \frac{1}{4M/N}$$

the argument of the sine function is less than $\pi/2$ and using the approximation $\sin(\theta) \approx \theta$ we find that $a[k]$ drops off as $1/k$, just as the partials of a sawtooth wave. But for larger values of k the sine term oscillates between 1 and -1, so that the amplitudes drop off irregularly as $1/k^2$.

Figure 10.8 shows the partial strengths with M/N set to 0.03; here, our prediction is that the $1/k$ dependence should extend to $k \approx 1/(4 \cdot 0.03) \approx 8.5$, in rough agreement with the figure.

Another way to see why the partials should behave as $1/k$ for low values of k and $1/k^2$ thereafter, is to compare the period of a given partial with the length of the short segment, $2M$. For partials numbering less than $N/4M$, the period is at least twice the length of the short segment, and at that scale the waveform is nearly indistinguishable from a sawtooth wave. For partials numbering in excess of $N/2M$, the two corners of the triangle wave are at least one period apart, and at these higher frequencies the two corners (each with $1/k^2$ frequency dependence) are resolved from each other. In the figure, the notch at partial 17 occurs at the wavelength $N/2M \approx 1/17$, at which wavelength the two corners are one cycle apart; since the corners are opposite in sign they cancel each other.

10.4 Predicting and Controlling Foldover

Now we descend to the real situation, in which the period of the waveform cannot be assumed to be arbitrarily long and integer-valued. Suppose (for definiteness) we want to synthesize tones at 440 Hertz (A above middle C), and that we are using a sample rate of 44100 Hertz, so that the period is about 100.25 samples. Theoretically, given a very high sample rate, we would expect the fiftieth partial to have magnitude 1/50 compared to the

fundamental and a frequency about 20 kHz. If we sample this waveform at
the (lower) sample rate of 44100, then partials in excess of this frequency
will be aliased, as described in Section 3.1. The relative strength of the
folded-over partials will be on the order of -32 decibels—quite audible. If
the fundamental frequency is raised further, more and louder partials reach
the Nyquist frequency (half the sample rate) and begin to fold over.

Foldover problems are much less pronounced for waveforms with only
corners (instead of jumps) because of the faster dropoff of higher partial
frequencies; for instance, a symmetric triangle wave at 440 Hertz would
get twice the dropoff, or -64 decibels. In general, though, waveforms with
discontinuities are a better starting point for subtractive synthesis (the most
popular classical technique). In case you were hoping, subtractive filtering
can't remove foldover once it is present in an audio signal.

10.4.1   Over-sampling

As a first line of defense against foldover, we can synthesize the waveform
at a much higher sample rate, apply a low-pass filter whose cutoff frequency
is set to the Nyquist frequency (for the original sample rate), then down-
sample. For example, in the above scenario (44100 sample rate, 440 Hertz
tone) we could generate the sawtooth at a sample rate of $16 \cdot 44100 = 705600$
Hertz. We need only worry about frequencies in excess of $705600 - 20000 =$
685600 Hertz (so that they fold over into audible frequencies; foldover to
ultrasonic frequencies normally won't concern us) so the first problematic
partial is $685600/440 = 1558$, whose amplitude is -64dB relative to that of
the fundamental.

This attenuation degrades by 6 dB for every octave the fundamental
is raised, so that a 10 kHz. sawtooth only enjoys a 37 dB drop from the
fundamental to the loudest foldover partial. On the other hand, raising the
sample rate by an additional factor of two reduces foldover by the same
amount. If we really wish to get 60 decibels of foldover rejection—all the
way up to a 10 kHz. fundamental—we will have to over-sample by a factor
of 256, to a sample rate of about 11 million Hertz.

10.4.2   Sneaky triangle waves

For low fundamental frequencies, over-sampling is an easy way to get adequate foldover protection. If we wish to allow higher frequencies, we will
need a more sophisticated approach. One possibility is to replace discontinuities by ramps, or in other words, to replace component sawtooth waves
by triangle waves, as treated in Section 10.3.4, with values of M/N small

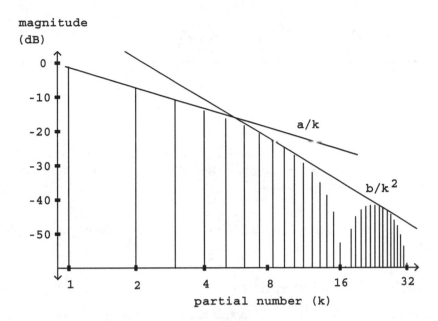

Figure 10.8: Magnitude spectrum of a triangle wave with $M/N = 0.03$. The two line segments show $1/k$ and $1/k^2$ behavior at low and high frequencies.

enough that the result sounds like a sawtooth wave, but large enough to control foldover.

Returning to Figure 10.8, suppose for example we imitate a sawtooth wave with a triangle wave with M equal to two samples, so that the first notch falls on the Nyquist frequency. Partials above the first notch (the 17th partial in the figure) will fold over; the worst of them is about 40 dB below the fundamental. On the other hand, the partial strengths start dropping faster than those of a true sawtooth wave at about half the Nyquist frequency. This is acceptable in some, but not all, situations.

The triangle wave strategy can be combined with over-sampling to improve the situation further. Again in the context of Figure 10.8, suppose we over-sample by a factor of 4, and set the first notch at the original sample rate. The partials up to the Nyquist frequency (partial 8, at the fundamental frequency shown in the figure) follow those of the true sawtooth wave fairly well. Foldover sets in only at partial number 48, and is 52 dB below the fundamental. This overall behavior holds for any fundamental frequency up to about one quarter the sample rate (after which M exceeds $N/2$). Setting the notch frequency to the original sample rate is equivalent

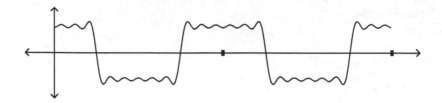

Figure 10.9: A square wave, band-limited to partials 1, 3, 5, 7, 9, and 11. This can be regarded approximately as a series of band-limited step functions arranged end to end.

to setting the segment of length $2M$ to one sample (at the original sample rate).

10.4.3   Transition splicing

In the point of view developed in this chapter, the energy of the spectral components of classical waves can be attributed entirely to their jumps and corners. This is artificial, of course: the energy really emanates from the entire waveform. Our derivation of the spectrum of the classical waveforms uses the jumps and corners as a bookkeeping device, and this is possible because the entire waveform is determined by their positions and magnitudes.

Taking this ruse even further, the problem of making band-limited versions of classical waveforms can be attacked by making band-limited versions of the jumps and corners. Since the jumps are the more serious foldover threat, we will focus on them here, although the approach described here works perfectly well for corners as well.

To construct a band-limited step function, all we have to do is add the Fourier components of a square wave, as many as we like, and then harvest the step function at any one of the jumps. Figure 10.9 shows the partial Fourier sum corresponding to a square wave, using partials 1, 3, 5, 7, 9, and 11. The cutoff frequency can be taken as 12ω (if ω is the fundamental frequency).

If we double the period of the square wave, to arrive at the same cutoff frequency, we would add twice as many Fourier partials, up to number 23, for instance. Extending this process forever, we would eventually see the ideal band-limited step function, twice per (arbitrarily long) period.

In practice we can do quite well using only the first two partials (one and three times the fundamental). Figure 10.10 (part a) shows a two-partial

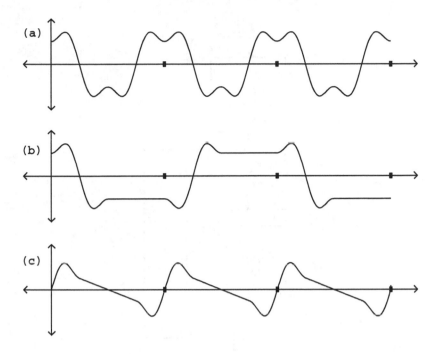

Figure 10.10: Stretching a band-limited square wave: (a) the original wave-form; (b) after splicing in horizontal segments; (c) using the same step transition for a sawtooth wave.

approximation of a square wave. The cutoff frequency is four times the fundamental; so if the period of the waveform is eight samples, the cutoff is at the Nyquist frequency. Part (b) of the figure shows how we could use this step function to synthesize, approximately, a square wave of twice the period. If the cutoff frequency is the Nyquist frequency, the period of the waveform of part (b) is 16 samples. Each transition lasts 4 samples, because the band-limited square wave has a period of eight samples.

We can make a band-limited sawtooth wave by adding the four-sample-long transition to a ramp function so that the end of the resulting function meets smoothly with itself end to end, as shown in part (c) of the figure. There is one transition per period, so the period must be at least four samples; the highest fundamental frequency we can synthesize this way is half the Nyquist frequency. For this or lower fundamental frequency, the foldover products all turn out to be at least 60 dB quieter than the fundamental.

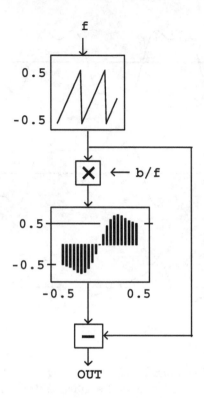

Figure 10.11: Block diagram for making a sawtooth wave with a spliced transition.

Figure 10.11 shows how to generate a sawtooth wave with a spliced transition. The two parameters are f, the fundamental frequency, and b, the band limit, assumed to be at least as large as f. We start with a digital sawtooth wave (a phasor) ranging from -0.5 to 0.5 in value. The transition will take place at the middle of the cycle, when the phasor crosses 0. The wavetable is traversed in a constant amount of time, $1/b$, regardless of f. The table lookup is taken to be non-wraparound, so that inputs out of range output either -0.5 or 0.5.

At the end of the cycle the phasor discontinuously jumps from -0.5 to 0.5, but the output of the transition table jumps an equal and opposite amount, so the result is continuous. During the portion of the waveform in which the transition table is read at one or the other end-point, the output describes a straight line segment.

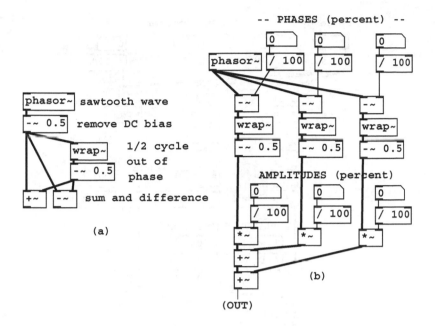

Figure 10.12: Combining sawtooth waves: (a) adding and subtracting saw-tooth waves 1/2 cycle out of phase, to extract even and odd harmonics; (b) combining three sawtooth waves with arbitrary amplitudes and phases.

10.5 Examples

Combining sawtooth waves

Example J01.even.odd.pd (Figure 10.12, part a) shows how to combine sawtooth waves in pairs to extract the even and odd harmonics. The resulting waveforms are as shown in Figure 10.3. Example J02.trapezoids.pd (part b of the figure) demonstrates combining three sawtooth waves at arbitrary phases and amplitudes; the resulting classic waveform has up to three jumps and no corners. The three line segments are horizontal as long as the three jumps add to zero; otherwise the segments are sloped to make up for the the unbalanced jumps so that the result repeats from one period to the next.

Example J03.pulse.width.mod.pd (not shown) combines two sawtooth waves, of opposite sign, with slightly different frequencies so that the relative phase changes continuously. Their sum is a rectangle wave whose width varies in time. This is known as pulse width modulation ("PWM").

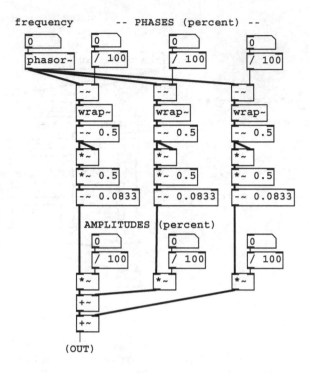

Figure 10.13: Combining parabolic waves to make a waveform with three corners.

Example J04.corners.pd (Figure 10.13) shows how to add parabolic waves to make a combined waveform with three corners. Each parabolic wave is computed from a sawtooth wave (ranging from -0.5 to 0.5) by squaring it, multiplying by 0.5, and subtracting the DC component of -1/12, or -0.08333. The patch combines three such parabolic waves with controllable amplitudes and phases. As long as the amplitudes sum to zero, the resulting waveform consists of line segments, whose corners are located according to the three phases and have slope changes according to the three amplitudes.

Strategies for band-limiting sawtooth waves

Example J05.triangle.pd (Figure 10.14, part a) shows a simple way to make a triangle wave, in which only the slope of the rising and falling segment are specified. A phasor supplies the rising shape (its amplitude being the slope), and the same phasor, subtracted from one, gives the decaying shape. The minimum of the two linear functions follows the rising phasor up to

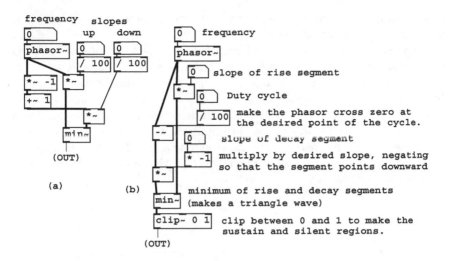

Figure 10.14: Alternative techniques for making waveforms with corners: (a) a triangle wave as the minimum of two line segments; (b) clipping a triangle wave to make an "envelope".

the intersection of the two, and then follows the falling phasor back down to zero at the end of the cycle.

A triangle wave can be clipped above and below to make a trapezoidal wave, which can be used either as an audio-frequency pulse or, at a lower fundamental frequency, as a repeating ASR (attack/sustain/release) envelope. Patch J06.enveloping.pd (Figure 10.14 part b) demonstrates this. The same rising shape is used as in the previous example, and the falling shape differs only in that its phase is set so that it falls to zero at a controllable point (not necessarily at the end of the cycle as before). The clip~ object prevents it from rising above 1 (so that, if the intersection of the two segments is higher than one, we get a horizontal "sustain" segment), and also from falling below zero, so that once the falling shape reaches zero, the output is zero for the rest of the cycle.

Example J07.oversampling.pd shows how to use up-sampling to reduce foldover when using a phasor~ object as an audio sawtooth wave. A subpatch, running at 16 times the base sample rate, contains the phasor~ object and a three-pole, three-zero Butterworth filter to reduce the amplitudes of partials above the Nyquist frequency of the parent patch (running at the original sample rate) so that the output won't fold over when it is down-sampled at the outlet~ object. Example J08.classicsynth.pd demonstrates using up-sampled phasors as signal generators to make an imitation of a classic synthesizer doing subtractive synthesis.

Example J09.bandlimited.pd shows how to use transition splicing as an alternative way to generate a sawtooth wave with controllable foldover. This has the advantage of being more direct (and usually less compute-intensive) than the up-sampling method. On the other hand, this technique depends on using the reciprocal of the fundamental frequency as an audio signal in its own right (to control the amplitude of the sweeping signal that reads the transition table) and, in the same way as for the PAF technique of Chapter 6, care must be taken to avoid clicks if the fundamental frequency changes discontinuously.

Exercises

1. A `phasor~` object has a frequency of 441 Hertz (at a sample rate of 44100). What is the amplitude of the DC component? The fundamental? The partial at 22050 Hertz (above which the partials fold over)?

2. A square wave oscillates between 1 and -1. What is its RMS amplitude?

3. In Section 10.3 a square wave was presented as an odd waveform whose Fourier series consisted of sine (and not cosine) functions. If the square wave is advanced 1/8 cycle in phase, so that it appears as an even function, what does its Fourier series become?

4. A rectangle wave is 1 for 1/4 cycle, zero for 3/4 cycles. What are the strengths of its harmonics at 0, 1, 2, 3, and 4 times the fundamental?

5. How much is $1 + 1/9 + 1/25 + 1/49 + 1/81 + \cdots$?

Bibliography

[Bal03]   Mark Ballora. *Essentials of Music Technology*. Prentice Hall, Upper Saddle River, New Jersey, 2003.

[Ble01]   Barry Blesser.   An interdisciplinary synthesis of reverberation viewpoints.   *Journal of the Audio Engineering Society*, 49(10):867–903, 2001.

[Bou00]   Richard Boulanger, editor. *The Csound book*. MIT Press, Cambridge, Massachusetts, 2000.

[Cha80]   Hal Chamberlin. *Musical applications of microprocessors*. Hayden, Rochelle Park, N.J., 1980.

[Cho73]   John Chowning.   The synthesis of complex audio spectra by means of frequency modulation. *Journal of the Audio Engineering Society*, 21(7):526–534, 1973.

[Cho89]   John Chowning. Frequency modulation synthesis of the singing voice. In Max V. Mathews and John R. Pierce, editors, *Current Directions in Computer Music Research*, pages 57–64. MIT Press, Cambridge, 1989.

[DJ85]   Charles Dodge and Thomas A. Jerse. *Computer music : synthesis, composition, and performance*. Schirmer, New York, 1985.

[DL97]   Mark Dolson and Jean Laroche. About this phasiness business. In *Proceedings of the International Computer Music Conference*, pages 55–58, Ann Arbor, 1997. International Computer Music Association.

[GM77]   John M. Grey and James A. Moorer. Perceptual evaluations of synthesized musical instrument tones. *Journal of the Acoustical Society of America*, 62:454–462, 1977.

323

[Har87]   William M. Hartmann. Digital waveform generation by fractional addressing. *Journal of the Acoustical Society of America*, 82:1883–1891, 1987.

[KS83]   Kevin Karplus and Alex Strong. Digital synthesis of plucked-string and drum timbres. *Computer Music Journal*, 7(2):43–55, 1983.

[Leb77]   Marc Lebrun. A derivation of the spectrum of FM with a complex modulating wave. *Computer Music Journal*, 1(4):51–52, 1977.

[Leb79]   Marc Lebrun. Digital waveshaping synthesis. *Journal of the Audio Engineering Society*, 27(4):250–266, 1979.

[Mat69]   Max V. Mathews. *The Technology of Computer Music*. MIT Press, Cambridge, Massachusetts, 1969.

[Moo90]   F. Richard Moore. *Elements of Computer Music*. Prentice Hall, Englewood Cliffs, second edition, 1990.

[PB87]   T. W. Parks and C.S. Burrus. *Digital filter design*. Wiley, New York, 1987.

[Puc95a] Miller S. Puckette. Formant-based audio synthesis using nonlinear distortion. *Journal of the Audio Engineering Society*, 43(1):224–227, 1995.

[Puc95b] Miller S. Puckette. Phase-locked vocoder. In *IEEE ASSP Workshop on Applications of Signal Processing to Audio and Acoustics*, 1995.

[Puc01]   Miller S. Puckette. Synthesizing sounds with specified, time-varying spectra. In *Proceedings of the International Computer Music Conference*, pages 361–364, Ann Arbor, 2001. International Computer Music Association.

[Puc05]   Miller S. Puckette. Phase bashing for sample-based formant synthesis. In *Proceedings of the International Computer Music Conference*, pages 733–736, Ann Arbor, 2005. International Computer Music Association.

[Reg93]   Phillip A. Regalia. Special filter design. In Sanjit K. Mitra and James F. Kaiser, editors, *Handbook for digital signal processing*, pages 907–978. Wiley, New York, 1993.

[RM69]   Jean-Claude Risset and Max V. Mathews. Analysis of musical instrument tones. *Physics Today*, 22:23–40, 1969.

[RMW02]   Thomas D. Rossing, F. Richard Moore, and Paul A. Wheeler. *The Science of Sound*. Addison Wesley, San Francisco, third edition, 2002.

[Roa01]   Curtis Roads. *Microsound*. MIT Press, Cambridge, Massachusetts, 2001.

[Sch77]   Bill Schottstaedt. Simulation of natural instrument tones using frequency modulation with a complex modulating wave. *Computer Music Journal*, 1(4):46–50, 1977.

[SI03]   Julius Orion Smith III. *Mathematics of the Discrete Fourier Transform (DFT), with Music and Audio Applications*. W3K Publishing, Menlo Park, California, 2003.

[Ste96]   Kenneth Steiglitz. *A Digital Signal Processing Primer*. Addison-Wesley, Menlo Park, California, 1996.

[Str85]   John Strawn, editor. *Digital Audio Signal Processing*. William Kaufmann, Los Altos, California, 1985.

[Str95]   Allen Strange. *Electronic music: systems, techniques, and controls*. W. C. Brown, Dubuque, Iowa, 1995.

Index

CPSIA information can be obtained
at www.ICGtesting.com
Printed in the USA
LVHW081634061222
734339LV00027B/294